Lecture Notes in Physics

For information about Vols. 1–90, please contact your bookseller or Springer-Verlag.

Vol. 91: Computing Methods in Applied Sciences and Engineering, 1977, II. Proceedings, 1977. Edited by R. Glowinski and J. L. Lions. VI, 359 pages. 1979.

Vol. 92: Nuclear Interactions. Proceedings, 1978. Edited by B. A. Robson. XXIV, 507 pages. 1979.

Vol. 93: Stochastic Behavior in Classical and Quantum Hamiltonian Systems. Proceedings, 1977. Edited by G. Casati and J. Ford. VI, 375 pages. 1979.

Vol. 94: Group Theoretical Methods in Physics. Proceedings, 1978. Edited by W. Beiglböck, A. Böhm and E. Takasugi. XIII, 540 pages. 1979.

Vol. 95: Quasi One-Dimensional Conductors I. Proceedings, 1978. Edited by S. Barišić, A. Bjeliš, J. R. Cooper and B. Leontič. X, 371 pages. 1979.

Vol. 96: Quasi One-Dimensional Conductors II. Proceedings 1978. Edited by S. Barišić, A. Bjeliš, J. R. Cooper and B. Leontić. XII, 461 pages. 1979.

Vol. 97: Hughston, Twistors and Particles. VIII, 153 pages. 1979.

Vol. 98: Nonlinear Problems in Theoretical Physics. Proceedings, 1978. Edited by A. F. Rañada. X, 216 pages. 1979.

Vol. 99: M. Drieschner, Voraussage – Wahrscheinlichkeit – Objekt. XI, 308 Seiten. 1979.

Vol. 100: Einstein Symposion Berlin. Proceedings 1979. Edited by H. Nelkowski et al. VIII, 550 pages. 1979.

Vol. 101: A. Martin-Löf, Statistical Mechanics and the Foundations of Thermodynamics. V, 120 pages. 1979.

Vol. 102: H. Hora, Nonlinear Plasma Dynamics at Laser Irradiation. VIII, 242 pages. 1979.

Vol. 103: P. A. Martin, Modèles en Mécanique Statistique des Processus Irréversibles. IV, 134 pages. 1979.

Vol. 104: Dynamical Critical Phenomena and Related Topics. Proceedings, 1979. Edited by Ch. P. Enz. XII, 390 pages. 1979.

Vol. 105: Dynamics and Instability of Fluid Interfaces. Proceedings, 1978. Edited by T. S. Sørensen. V, 315 pages. 1979.

Vol. 106: Feynman Path Integrals, Proceedings, 1978. Edited by S. Albeverio et al. XI, 451 pages. 1979.

Vol. 107: J. Kijowski, W. M. Tulczyjew, A Symplectic Framework for Field Theories. IV, 257 pages. 1979.

Vol. 108: Nuclear Physics with Electromagnetic Interactions. Proceedings, 1979. Edited by H. Arenhövel and D. Drechsel. IX, 509 pages. 1979.

Vol. 109: Physics of the Expanding Universe. Proceedings, 1978. Edited by M. Demiański. V, 210 pages. 1979.

Vol. 110: D. A. Park, Classical Dynamics and Its Quantum Analogues. VIII, 339 pages. 1979.

Vol. 111: H.-J. Schmidt, Axiomatic Characterization of Physical Geometry. V, 163 pages. 1979.

Vol. 112: Imaging Processes and Coherence in Physics. Proceedings, 1979. Edited by M. Schlenker et al. XIX, 577 pages. 1980.

Vol. 113: Recent Advances in the Quantum Theory of Polymers. Proceedings 1979. Edited by J.-M. André et al. V, 306 pages. 1980.

Vol. 114: Stellar Turbulence. Proceedings, 1979. Edited by D. F. Gray and J. L. Linsky. IX, 308 pages. 1980.

Vol. 115: Modern Trends in the Theory of Condensed Matter. Proceedings, 1979. Edited by A. Pekalski and J. A. Przystawa. IX, 597 pages. 1980.

Vol. 116: Mathematical Problems in Theoretical Physics. Proceedings, 1979. Edited by K. Osterwalder. VIII, 412 pages. 1980.

Vol. 117: Deep-Inelastic and Fusion Reactions with Heavy Ions. Proceedings, 1979. Edited by W. von Oertzen. XIII, 394 pages. 1980.

Vol. 118: Quantum Chromodynamics. Proceedings, 1979. Edited by J. L. Alonso and R. Tarrach. IX, 424 pages. 1980.

Vol. 119: Nuclear Spectroscopy. Proceedings, 1979. Edited by G. F. Bertsch and D. Kurath. VII, 250 pages. 1980.

Vol. 120: Nonlinear Evolution Equations and Dynamical Systems. Proceedings, 1979. Edited by M. Boiti, F. Pempinelli and G. Soliani. VI, 368 pages. 1980.

Vol. 121: F. W. Wiegel, Fluid Flow Through Porous Macromolecular Systems. V, 102 pages. 1980.

Vol. 122: New Developments in Semiconductor Physics. Proceedings, 1979. Edited by F. Beleznay et al. V, 276 pages. 1980.

Vol. 123: D. H. Mayer, The Ruelle-Araki Transfer Operator in Classical Statistical Mechanics. VIII, 154 pages. 1980.

Vol. 124: Gravitational Radiation, Collapsed Objects and Exact Solutions. Proceedings, 1979. Edited by C. Edwards. VI, 487 pages. 1980.

Vol. 125: Nonradial and Nonlinear Stellar Pulsation. Proceedings, 1980. Edited by H. A. Hill and W. A. Dziembowski. VIII, 497 pages. 1980.

Vol. 126: Complex Analysis, Microlocal Calculus and Relativistic Quantum Theory. Proceedings, 1979. Edited by D. Iagolnitzer. VIII, 502 pages. 1980.

Vol. 127: E. Sanchez-Palencia, Non-Homogeneous Media and Vibration Theory. IX, 398 pages. 1980.

Vol. 128: Neutron Spin Echo. Proceedings, 1979. Edited by F. Mezei. VI, 253 pages. 1980.

Vol. 129: Geometrical and Topological Methods in Gauge Theories. Proceedings, 1979. Edited by J. Harnad and S. Shnider. VIII, 155 pages. 1980.

Vol. 130: Mathematical Methods and Applications of Scattering Theory. Proceedings, 1979. Edited by J. A. DeSanto, A. W. Sáenz and W. W. Zachary. XIII, 331 pages. 1980.

Vol. 131: H. C. Fogedby, Theoretical Aspects of Mainly Low Dimensional Magnetic Systems. XI, 163 pages. 1980.

Vol. 132: Systems Far from Equilibrium. Proceedings, 1980. Edited by L. Garrido. XV, 403 pages. 1980.

Vol. 133: Narrow Gap Semiconductors Physics and Applications. Proceedings, 1979. Edited by W. Zawadzki. X, 572 pages. 1980.

Vol. 134: $\gamma\gamma$ Collisions. Proceedings, 1980. Edited by G. Cochard and P. Kessler. XIII, 400 pages. 1980.

Vol. 135: Group Theoretical Methods in Physics. Proceedings, 1980. Edited by K. B. Wolf. XXVI, 629 pages. 1980.

Lecture Notes in Physics

Edited by J. Ehlers, München K. Hepp, Zürich
R. Kippenhahn, München H. A. Weidenmüller, Heidelberg
and J. Zittartz, Köln

160

Unified Theories of Elementary Particles

Critical Assessment and Prospects

Proceedings of the Heisenberg Symposium
Held in München, July 16–21, 1981

Edited by P. Breitenlohner and H.P. Dürr

Springer-Verlag
Berlin Heidelberg GmbH 1982

Editors
Hans-Peter Dürr
Peter Breitenlohner
Max-Planck-Institut für Physik und Astrophysik
Werner Heisenberg Institut für Physik
Föhringer Ring 6, D-8000 München 40

ISBN 978-3-540-11560-1 ISBN 978-3-540-39250-7 (eBook)
DOI 10.1007/978-3-540-39250-7

Originally published by Springer-Verlag Berlin Heidelberg New York in 1982

2153/3140-543210

Preface

Many attempts are presently being made to establish a common dynamical basis for the various forces and the particle structure observed in nature. The quest for a unified theory governing the dynamics of all material bodies has a very long tradition in this Institute.

Albert Einstein was appointed first director of this Institute in 1913, which then was called the Kaiser-Wilhelm-Institut für Physik and was situated in Berlin-Dahlem. In 1915 Einstein succeeded in formulating his General Theory of Relativity which led to a union of space-time geometry and gravitation. During the second half of his life he tried very hard to find appropriate generalizations to incorporate Maxwell's electrodynamics into the geometrical scheme as well. This ambitious attempt to arrive at a "einheitliche Feldtheorie" of all the forces known at that time ultimately failed for essentially two reasons. First, Maxwell's electrodynamics in the course of time - and, in fact, as an immediate consequence of Einstein's own famous paper on the photoelectric effect in 1905 - changed from a classical theory to a quantum theory, into quantum electrodynamics. Secondly, new types of forces were discovered in probing the atomic nuclei, the hadronic and weak interactions, which could only be accommodated into the framework of quantum field theories. Einstein's generalizations on his General Theory of Relativity did not allow accounting for the quantum structure and did not provide the proper features for a geometrical interpretation of the new forces.

It was Werner Heisenberg who was the first to attempt a unification of all forces within the framework of a quantum field theory and elementary particle physics. For nearly three decades he was the director of this Institute, first in Berlin, then in Göttingen and finally in Munich, until his retirement in 1970. His interest in the unification or rather in the formulation of a fundamental law underlying the dynamics of elementary particles goes back to the late thirties, stimulated by observations of the multiple production of elementary particles in cosmic rays.

We have proclaimed this Symposium in memory of Werner Heisenberg who died five years ago. This year, on December 5th, we will celebrate his 80th birthday, on which occasion, according to a decree of the Max Planck Society, this Institute will recieve the additional name "Werner-Heisenberg-Institut".

The two different approaches to unification of the general dynamics in nature - the classical geometrical approach and the quantum mechanical elementary particle approach - will be the main topics of this Symposium. It appears, at first, that both approaches have very little in common because they are established on different levels which makes a direct comparison difficult. In the course of time, however, we have learned that, despite their different points of departure, there are many interesting points of contact. Important links are provided by the gauge features of the fundamental particle interactions which allow a geometrical interpretation, and the nontrivial properties of the quantum-mechanical ground-state and soliton-type solutions which exhibit classical aspects in a purely quantum mechanical description. One of the goals of the Symposium was to emphasize these points of contact and to stimulate looking at the same problem from different angles.

Besides the plenary lectures presented in this volume, two discussion sessions were arranged during the Symposium on the general topics: "Unification and Subquark Schemes" and "Geometrical Approaches to Unification". These discussion sessions - with the plenary lecturers serving as a panel - aimed at a critical assessment of the presented theories and an estimation of their future prospects. Unfortunately, these lively and articulate discussions were not recorded and hence could not be included in the proceedings.

The editors would like to take the opportunity to express their sincere gratitude to all the speakers who have so spontaneously agreed to participate in this Symposium. Their continued readiness to frankly express their hopes and fears has helped greatly to reach a better understanding of the subject matter. The editors would also like to thank them for their cooperation in preparing the manuscripts for these proceedings.

The Heisenberg Symposium would not have been possible without generous financial support from the Max Planck Society and the assistance of the Organizing Committee and many members of the Max-Planck-Institut für Physik. This is gratefully acknowledged.

München, September 1981

P. Breitenlohner H.P. Dürr

Organizing Committee:

W. Drechsler
D. Maison
R.D. Peccei
H. Rechenberg
H. Saller
E. Seiler
F. Wagner

Conference Secretaries:

W. Huber
I. Kraus
C. Maseko
J. Peccei

TABLE OF CONTENTS

AN APPROACH TO THE UNIFICATION OF ELEMENTARY PARTICLE INTERACTIONS

Mary K. Gaillard
Lawrence Berkeley Laboratory
and
Department of Physics
University of California
Berkeley, California 94720 U.S.A.

I. INTRODUCTION

Grand unified theories[1,2,3] (GUTs) have met with a few rather impressive successes. First, they are experimentally viable in that the assumption that the three coupling constants of the strong and electroweak gauge theories become equal at a common renormalization scale is consistant[4,5,3] with the measured value of the weak angle, and under more restrictive assumptions,[2] with the b-quark to τ-lepton mass ratio.[6,5,7] Secondly they predict the instability of matter, giving an estimated nucleon life-time that should be measurable in the near future, and, for the first time, the possibility of understanding the cosmological predominance of matter over anti-matter. In addition, these theories ensure the quantization of electric charge.

While possessing many attractive features, grand unified theories are, in the view of most theorists, clearly incomplete. The first obvious defect is the proliferation of arbitrary parameters. Among these are: the Yukawa couplings which determine the fermion mass spectrum and flavor mixing in weak decays; the scalar self-couplings which determine the pattern of symmetry breaking which in turn determines the masses of vector mesons associated with broken gauge symmetries. Both the Yukawa and scalar couplings may play a role in the CP violating parameters of the resultant broken-symmetric theory. A unified gauge theory as such possesses no criterion for fixing its fermion and scalar content, nor for the initial choice of gauge group. The second notorious difficulty is the infamous "gauge hierarchy problem"; this is really a specific case of the above-mentioned arbitrariness, but it is particularly acute in that it entails the understanding of a ratio of mass scales which differ by many orders of magnitude, or -equivalently- by many powers of the coupling constant. Finally the force of gravity is not included in our present picture of unified interactions. Our theories extrapolate in energy to about 13 orders of magnitude beyond presently observed energies while ignoring quantum gravitational effects which should become significant at an energy scale only four orders of magnitude larger than that presently accepted as the "unification energy".

It is clear that we need physical criteria which lie outside the scope of gauge theories in order to further restrict our model building. During the last decade the underlying criterion has been renormalizability. The reason why the scalar and spin 1/2 content and coupling constants remain arbitrary is that any interactions among these fields of dimension ≤ 4 gives a renormalizable theory. On the other hand we know how to construct a renormalizable theory of spin 1 fields only if it encorporates an exact or spontaneously broken gauge symmetry. This fixes the spectrum of vector fields — their multiplicity is given by the adjoint representation of the gauge group — and determines their self couplings in terms of a single arbitrary constant. Once the spin 1/2 and scalar content has been specified, their couplings to vector fields are also determined in terms of the same constant. For higher spins there is no known renormalizable theory, but supergravity[8] offers the promise of a highly convergent theory for spin ≤ 2. For spin ≥ 2 one does not even know how to write down a field theory.

If we wish to take a lesson from recent history, we may note the following. Physical observation, namely, the existence of spin 1 charged weak currents, together with the criterion of renormalizability led to the construction[9] of the experimentally successful electroweak gauge theory. An analogy might be that physical observation namely, the existence of gravitational interactions, requiring spin 2 in a quantum formulation, together with the criterion of a sufficiently convergent theory to allow for the calculation of S-matrix elements in terms of a limited number of input parameters (only the gravitational constant κ?) would lead uniquely to supergravity as the theory of fundamental interactions. Extended supergravity, which embeds internal symmetries as well as supersymmetries,[10] may determine uniquely the elementary particle content of the theory.

The above conjecture leads naturally to the question as to whether the gauge theories which apparently describe well observed phenomena can be embedded into an extended supergravity theory.[11] Extended supersymmetry is characterized by a number N which specifies both the number of independent supersymmetry transformations and the number of degrees of freedom associated with internal symmetries. For a massless supermultiplet, the helicities λ run over the range [12]

$$J - \frac{N}{2} \leq \lambda \leq J \ . \tag{1}$$

If we restrict ourselves to known field theories we must impose

$$|\lambda| \leq 2 \quad \text{or} \quad N \leq 8 \ . \tag{2}$$

This constraint restricts the possible supergravity theories to those of Table 1. A theory which is symmetric under N supersymmetries possesses[13] a rigid (usually called global) U(N) symmetry except for $N = 8$, in which case the basic (unique for $|\lambda| \leq 2$) supermultiplet is self-conjugate, and does not admit a U(1) symmetry ($N = 7$ and $N = 8$ theories have the same spectra and are believed to be the same theory). However we are interested in gauge symmetries. In conventional theories these require vector fields in the adjoint representation. Within this framework supergravity theories allow at most the gauging of an orthogonal group SO(N), $N \leq 8$, which is not large enough to

TABLE I. Possible supergravity theories

N	Rigid symmetry	Number of vector fields	Possible gauge symmetry
$\leqslant 5$	U(N)	$\frac{1}{2}N(N-1)$	SO(N)
6	U(6)	16	SO(6) × U(1)
7,8	SU(8)	28	SO(8)

accomodate phenomenologically acceptable grand unified gauge theories for which the minimal gauge group is SU(5). In fact SO(8) does not even contain the "observed" broken gauge group $SU(3)_c \times SU(2)_L \times U(1)$ which is believed to describe the strong and electroweak interactions. On the other hand, it does contain[11] the group $SU(3)_c \times U(1)_{em}$ of the presumably exact strong and electromagnetic gauge symmetries (along with an additional U(1) factor) and a possible point of view would be that only the exact gauge symmetries are described by fundamental gauge couplings while the others are somehow dynamically generated. This of course runs counter to the whole idea of grand unification and renders the above mentioned successes fortuitous. In addition one finds[11] that the observed quarks and leptons cannot all be accomodated in the spectrum of elementary particles, which is unique for N = 8 supergravity.

One is thus led to the idea of a composite structure for at least some fields of the conventional theories. Once this idea is accepted we shall see that not only is the allowed particle spectrum enlarged, but the possible gauge group is also enlarged in such a way that the conventional grand unification approach remains intact, except that the "observed" gauge interactions emerge as the effective couplings of composite states, and the grand unification group is no longer arbitrary.

II. HIDDEN SYMMETRIES OF SUPERGRAVITY

The idea of embedding observed gauge symmetries within an extended supergravity theory was given renewed impetus by the analysis of Cremmer and Julia[14] who showed that extended supergravity theories for $4 \leqslant N \leqslant 8$ possess an invariance under a group which is the direct product of a rigid non-compact group $\mathcal{G}$ with a local compact group $\mathcal{K}$, where $\mathcal{K}$ is U(N) for N = 4, 5, 6 and SU(8) for N = 7, 8. The scalar fields of the theory are valued on the coset space $\mathcal{G}/\mathcal{K}$. For N = 8 supergravity the invariance group is

$$\mathcal{G} \times \mathcal{K} = E_{7(+7)} \times SU(8) \quad . \tag{2}$$

In a manifestly gauge invariant formulation, there are 133 scalars, corresponding to the adjoint representation of E_7; these decompose into 63 (adjoint of SU(8)) which can be removed by a gauge transformation and 70 (coset space $\mathcal{G}/\mathcal{K}$) which are the physical scalars of the "special" or "physical" gauge. Expressed in terms of physical fields alone, with only 70 scalars, the theory has no manifest SU(8) gauge invariance but only a rigid SU(8) invariance, corresponding to the compact part of $E_{7(+7)}$, while the non-compact part of E_7 is realized non-linearly with parameters which depend on the scalar fields. The full $E_7 \times SU(8)$ invariance becomes manifest in a general gauge where the 63 unphysical scalars are present; gauge invariance is implemented in the usual way through the covariant derivative

$$D_\mu = \partial_\mu - Q_\mu \qquad (3)$$

where Q_μ represents a set of vector fields transforming as an SU(8) adjoint, for which, however, there is no kinetic energy term. The equations of motion then determine the Q_μ as composite (bilinear and higher order) operators in terms of the elementary fields of the N = 8 supermultiplet. At first sight SU(8) gauge invariance appears to be an artifact of the general gauge formulation which disappears when the theory is expressed in terms of physical particles. However, Cremmer and Julia conjectured that the two-point function for Q_μ developes a pole at the origin, resulting in massless spin-one states which might be identified with the gauge vectors of the "observed" theory. They supported their conjecture by analogy with CP^{n-1} models in two dimensions where a similar effect has been found[15] to occur in the 1/n expansion.

An immediate consequence of the Cremmer-Julia conjecture is that the grand unified theory (excluding generation unification) is necessarily the "minimal" one, i.e. the Georgi-Glashow[2] SU(5). The larger GUTs which are phenomenologically acceptable such as SO(10) and (compact) E_6 cannot be embedded in SU(8).

A second consequence is that, since the Q_μ necessarily belong to a supermultiplet of fields, one is naturally led to the EGMZ conjecture[16, 17] that the superpartners of the Q_μ also develope poles at the origin, a conjecture which is similarly supported by analogy with supersymmetric CP^n models.[18] In this way we would obtain a set of composite, massless states which should include all the fermions, vectors and scalars of the "observed" gauge theory. In this framework the Lagrangian of the "observed" theory, by which we mean the fully "grand" unified theory (i.e. SU(5)), is viewed as an effective Lagrangian describing the interactions of composite states which can be treated as local interactions of elementary fields at energies well below the Planck mass m_P, or equivalently over distances large compared with the Planck length $\kappa = m_P^{-1}$.

A problem immediately arises. Any N = 8 supermultiplet contains spins at least as large as 2, while a renormalizable field theory can have maximum spin 1. In order for an effective renormalizable theory to emerge at the grand unification mass $m_{GU} \sim 10^{-4} m_P$, supersymmetry must be broken at a scale $\gg m_{GU}$, perhaps already at m_P. For this we appeal to some unknown dynamical mechanism, and to "Veltman's theorem":[19]

> Given a set of bound states of radius r and mass $m \ll r^{-1}$, their couplings at energy scales $E \ll r^{-1}$ can be described by an effective local perturbation theory only if they are renormalizable.

This is not really a theorem,[20] but rather a statement of self consistancy. A perturbative treatment of non-renormalizable interactions requires the introduction of a cut-off Λ for which the only scale available is $\Lambda \sim r^{-1} \sim \kappa^{-1}$ in our case. It could be this highly divergent behavior which triggers the breaking of supersymmetry in such a way that a sensible low mass effective theory remains. This seems to be the conslusion that one is inevitably led to if N = 8 supergravity is the underlying theory, since experiment tells us that quarks and leptons, at least some of which are necessarily bound states with $m \ll \kappa^{-1}$, have interactions which are successfully described by a local, renormalizable perturbation theory at energy scales $E \ll m_P$.

Accepting the above set of conjectures we must address ourselves to two major questions.

1. What good is supersymmetry if it is so badly broken? For one thing it may determine[16] the particle spectrum just as the hadron spectrum is determined in terms of flavor SU(F) which is badly broken by quark mass differences. Secondly it may relate [21] the so-far unconstrained parameters of the GUT, just as badly broken SU(5) relates otherwise unconstrained parameters, such as $\sin^2\theta_W$ and m_b/m_τ, of the "low energy" $SU(3)_c \times SU(2)_L \times U(1)$ gauge theory. It is also possible that part of the supersymmetry remains unbroken down to a much lower energy scale, a possibility which could be relevant to the gauge hierarchy problem.[22]

2. What happened to all the other states of the bound supermultiplet? Possibilities include:[23] they acquire masses $O(m_P)$; their couplings are suppressed by powers of κ; they never bound in the first place.

In the remainder of this talk I will review the progress, or lack of it, on confronting these issues.

III. THE SPECTRUM OF COMPOSITE STATES

The EGMZ hypothesis[16] is that the full particle content of the grand unified theory should be found among the states of the irreducible massless supermultiplet which includes the zero-mass-shell projection of the composite SU(N) vector operators $\mathcal{Q}_\mu$ of $4 \leq N \leq 8$ supergravity. Various arguments[16,24,25] suggest that the appropriate multiplet has the structure

$$(-3/2)^A,\ (-1)^A_B,\ (-1/2)^A_{[BC]},\ (0)^A_{[BCD]},\ (+1/2)^A_{[BCDE]}\ \ldots.(5/2)^A \qquad (4)$$

+ T.C.P. conjugate states,

where A, B,... = 1, ..., N are SU(N) indices, brackets denote total antisymmetrization, and helicity is indicated in parenthesis.[12] The first remark to be made is that the spectrum (4) contains the usual three generations of fermions only for N = 8. The helicity and SU(8) content for this case is displayed in Table 2 where for $-1 \leqslant \lambda \leqslant 2$ the smaller representation is obtained by saturating the upper index with one lower index in the tensors (4), and the larger representation is the traceless part. The states enclosed by a dashed line correspond to the possible

TABLE II. Spectrum of composite states for N = 8 (TCP conjugates understood)

helicity	- 3/2	- 1	- 1/2	0	1/2	1	3/2	2	5/2
SU(8)	$\overline{8}$	63	216	420	$\overline{504}$	$\overline{378}$	$\overline{168}$	$\overline{36}$	$\overline{8}$
content		1	8	28	56	70	$\overline{56}$	$\overline{28}$	

content of a renormalizable gauge theory. However it must be further restricted so that the surviving fermion states form an anomaly-free set with respect to the surviving gauge group. This immediately rules out SU(8) as the effective renormalizable gauge theory. If we further impose the observational constraint that left and right handed fermions transform according to inequivalent (chiral) representations of the GUT, but equivalent (vector-like) representations under $SU(3)_c \times U(1)_{em}$ we find[21] that SU(6) and SU(7) are also eliminated as candidate effective theories. Since SU(5) is phenomenologically the minimal GUT group we are led uniquely to this choice. We note in addition[21] that if the 420 of scalars acquires a vacuum expectation value of order m_P, the natural scale of the theory, the largest surviving simple gauge group is again SU(5). Finally, we observe that if (why? in analogy with the retention of only the 63 of spin 1?) we retain only the 216 + $\overline{504}$ of left-handed ($\lambda = -1/2$) fermions, we find that the maximal subset which is anomaly free under SU(5) is

$$(45 + \overline{45}) + 4(24) + 9(10 + \overline{10}) + 3(5 + \overline{5}) + 9(1) + 3(\overline{5} + 10) \quad . \tag{5}$$

The $3(\overline{5} + 10)$ can then be identified with the "observed" three generations of light fermions, while the remaining states can acquire SU(5) invariant Dirac or Majorana masses which are expected to be of order m_{GU} or larger. In addition to the fermions (5) and their right-handed charge conjugate states the particle content of the effective GUT will include the 24 gauge vectors of SU(5) and some (or all) of the scalars of Table II. We note that the scalar content is sufficiently rich to satisfy any phenomenological needs of conventional SU(5).

IV. SCENARIOS FOR THE MISSING STATES

In this section I shall elaborate somewhat on the possibilities mentioned above for disposing of those states in Table II which cannot be embedded in the effective renormalizable GUT.

They acquire masses $O(m_P)$. In the present picture we have the following pattern of gauge symmetry breaking

$$SU(8) \xrightarrow[10^{19}\text{GeV}]{} SU(5) \xrightarrow[10^{15}\text{GeV}]{} SU(3)_c \times SU(2)_L \times U(1) \xrightarrow[10^{2}\text{GeV}]{} SU(3)_c \times U(1)_{em} \qquad (6)$$

where the first step at $m_P = 10^{19}$ GeV is attributed to dynamical effects and the subsequent breaking of SU(5) proceeds according to the conventional Higgs mechanism, with the Higgs scalars being suitable composites among those of Table II. Those masses which are generated at the first step, and which can be $O(m_P)$, must necessarily be SU(5) invariant. It is easy to see[16] that the high spin states of Table II cannot all obtain masses $O(m_P)$ in this way. It has further been found[16] that starting with any irreducible supermultiplet one cannot give masses to high spin states which are invariant even under the unbroken gauge group $SU(3)_c \times U(1)_{em}$. This could be achieved by disregarding conventional SU(5) and reinterpreting the quantum numbers of the basic multiplet so as to make the theory completely vector-like under $SU(3)_c \times U(1)_{em}$, but this cannot be done[16] in a way which reproduces the observed charge spectrum of fermions. There are [25] at least some sets of irreducible supermultiplets which allow SU(8) invariant masses, namely those which are obtained in the zero mass limit from an irreducible massive supermultiplet. However this set allows group invariant masses for all states and no light mass sector would be expected to survive. Attempts to find finite sets of supermultiplets other than the above type have given negative results.[26]

The remaining possibility[24] is that the surviving low mass spectrum is derived from an infinite set of supermultiplets. This idea has some support both from dynamical[27] and group theoretical[23,24] considerations. Recall that N = 8 supergravity is invariant under a non-compact E_7 which is realized non-linearly on the elementary fields (preons):

$$\psi \to F(\phi)\psi \qquad (7)$$

where ψ is a preon state of arbitrary spin and ϕ represents the 70-plet of scalar fields in the special gauge. If we wish to obtain an effective bound state theory where no preon fields appear, E_7 must be realized linearly, in which case the unitary representations are infinite. We may again appeal to an analogy with CP^n models with a non-compact group, where studies[28] suggest that the bound state spectrum forms a linear representation of the full group.

Consider a preon field ψ which transforms according to some representation R of SU(8). Then under an infinitesimal E_7 transformation of parameter ε

$$\delta\psi \sim \varepsilon\phi\psi \tag{8}$$

which transforms according to the reducible SU(8) representation

$$R' = R \times 70 \tag{9}$$

Now consider a bound state B which is the zero-mass-shell projection of an operator

$$\mathcal{O} = \psi_1\psi_2 + \text{higher order terms} \tag{10}$$

$\mathcal{O}$ and therefore B will transform according to some representation

$$R_B \in R_1 \times R_2 \tag{11}$$

of SU(8). From (9) we find

$$\delta\mathcal{O} \sim \varepsilon\psi_1\psi_2\phi + \ldots \tag{12}$$

so that the state B' obtained from an infinitesimal transformation on B

$$\delta B \sim \varepsilon B' \tag{13}$$

should transform according to a reducible representation

$$R_{B'} \in 70 \times R_B \quad . \tag{14}$$

A further transformation will generate a representation

$$R_{B''} \in (70 \times 70)_{sym} \times R_B \quad , \tag{15}$$

and so on.

From a general group theoretical analysis it is easy to see that for a given representation R of SU(8), the set of states

$$R_n = \{70\}^n \times R, \; n = 0, \ldots, \infty, \tag{16}$$

where $\{70\}^n$ represents a totally symmetrized product of 70's, is a representation of E_7. It is a reducible representation which is infinitely degenerate in each SU(8) representation which it contains. However if the binding dynamics respect both E_7 and supersymmetry, we must look for a set of states which represent the full algebra of symmetry generators. A more detailed analysis[23] shows that the set (16), where in our case the SU(8) representations R for each helicity are those of Table II, does represent the algebra and suggests that no reduction is possible. In this case we obtain[23] the following result: it is possible to give masses to any arbitrary set of the states (16), while keeping any arbitrary set massless in a way which is invariant under a subgroup of SU(8) which can be as large as SU(6). This result also holds if the set (16) is reduced only by limiting the multiplicity of each distinct SU(8) representation appearing in (16) to some finite (non-zero) value.

What is to be concluded from the above analysis? Simply that the scenario (6), with group invariant masses generated at each step, is a self-consistent possibility. However in this framework we have no criterion for determining which set of states survives at low energy, nor even for assuring that the low mass states all belong to our original supermultiplet of Eq. (4) and Table II, except that this is perhaps the most aesthetic hypothesis and the one which is most likely to constrain the effective GUT.

They have couplings $O(\kappa^n)$. S-matrix theorems[29] for threshold behavior imply that a state of helicity λ must couple with a dimensionful parameters

$$G_\lambda \propto m^{|2\lambda|-3} \tag{17}$$

where in our case we expect $m = \kappa^{-1}$. This implies that states of helicity $|\lambda| \geqslant 5/2$ have couplings $|G| \leqslant \kappa^2$ which are weaker than gravity and unobservable. States of helicity $|\lambda| = 2$ could have couplings of gravitational strength but their effect would be incoherent and unobservable for states which are not $SU(3)_c \times U(1)$ invariants; the others can have masses $\gtrsim O(m_W)$. Vector fields which are in a non-self-conjugate SU(8) representation have well defined transformation properties only when expressed in terms of the field strength and its dual, so that their SU(8) invariant couplings must again involve a dimensional constant $G \lesssim \kappa$. These arguments do not directly imply the decoupling of $|\lambda| = 3/2$ states nor of the $|\lambda| = 1/2$ states which are excluded from the anomaly free sector.

An alternative approach[30] is to disregard (as decoupled?) the high spin sector $|\lambda| \geqslant 1$ (except for the adjoint of vectors) and demand that the full $|\lambda| = 1/2$ spectrum form an anomaly free set under SU(8). Taking account of the selection rule[23] which requires that fermionic bound states of the supergravity preons must have an odd number of SU(8) indices, it is found that an even number of SU(5) generations of light $(\bar{5} + 10)$ fermions will occur, and that the initial bound state spectrum must include a repetition of some supermultiplets (to be interpreted as radial excitations?)

One might also ask whether there are any cosmological constraints limiting the number of decoupled massless helicity states. This turns out[23] not to be the case as long as the number of states which decoupled at the Planck temperature is sufficiently or smaller than the number of states remaining in equilibrium, which is the case for the set (5) and also for smaller sets which one might be led to consider[23] if one wished to retain an unbroken simple supersymmetry for the effective gauge theory.

They never bind. There are arguments[31] based on Lorentz covariance that massless states with helicity $|\lambda| \geqslant 1$ cannot carry a quantum number [e.g. SU(8), SU(5)] associated with a local vector current. Fortunately this argument is not applicable[32] in our case, as it would also exclude gauge vectors as bound states. The conserved SU(8) current[32] of N = 8 supergravity is not invariant under abelian gauge transformations

$$A_\mu \to A_\mu + \partial_\mu \lambda$$

on the preon vector fields A_μ. This means that the current can only be specified once a gauge is fixed, and that its Lorentz transformation properties are not well defined.

On the other hand, as discussed above, "Veltman's theorem " implies that supersymmetry must break dynamically at the Planck scale in order that an effective low energy theory may emerge. Perhaps this symmetry breaking is associated with the binding mechanism itself, preventing the formation of those states which would spoil renormalizability.

IV. CONCLUSIONS AND PROSPECTS

Ideally, the assumption that the grand unified theory emerges as an effective theory describing bound states of N = 8 supergravity preons should determine the GUT particle spectrum and constrain their couplings. Analysis of the spectrum has led to some possibly encouraging indications. At the least, the particle content in scalars, vectors and fermions needed to reproduce SU(5) phenomenology can be found among the states of the EGMZ multiplet.[16]

As for constraints on couplings, little progress has been made at present. One possibility is that some supersymmetry survives below the Planck mass. Renormalizability requires $|\lambda| \leqslant 1$, which from (1) implies $N \leqslant 4$ for the number of surviving supersymmetries. However, the empirical observation that fermions form chiral representations of the gauge group restricts this number to 1. The further study of supersymmetric GUTs, as well as of the non-renormalization properties of supersymmetric theories may prove enlightening, particularly with regard to the "gauge hierarchy" problem.

Electroweak phenomenology requires the Higgs scalar of the Glashow-Weinberg-Salam model[9] to have a mass smaller than about a TeV. On the other hand, when the electroweak model is embedded in a GUT, the scalar mass generally acquires a contribution of order 10^{15} GeV through its coupling to the 24-plet of scalars which generate the break-down of SU(5) to the $SU(3)_c \times SU(2)_L \times U(1)$ of GWS. In order to fit phenomenology, this contribution must be made to vanish by artificially adjusting coupling constant ratios or by cancelling it against equally large explicit mass terms. The required cancellation can indeed be arranged at any order in perturbation theory but it is highly unstable against radiative corrections which for n loops give contributions of order $(\alpha/4\pi)^n \times 10^{15}$ GeV, where $\alpha \cong 1/40$ is the GUT coupling constant.

A natural solution to the problem would occur if there were some symmetry which forced the mass of the GWS Higgs scalar to vanish. In gauge theories gauge invariance forces vectors to be massless and the chiral nature of the gauge couplings requires massless fermions. Since supersymmetry requires fermion-boson degeneracy, a supersymmetric chiral gauge theory would require massless scalars as well. Within this scenario supersymmetry must remain valid at energy scales down to 1 TeV or less. However this more or less obvious way of implementing the requirement $m_H \lesssim$ TeV turns out to have other phenomenological difficulties.

There is another property of supersymmetric theories which is not entirely understood: the so-called non-renormalization theorem[33] which ensures that if cancellations of the type needed to keep the Higgs mass small are imposed on the Lagrangian at the tree level, they will be stable against radiative corrections. SU(5) models[34] of this type yield satisfactory phenomenology if the symmetry breaking arises from terms of dimension two or less with parameter $m^2 \lesssim (1 \text{ TeV})^2$.

Theories of this type have observable consequences for proton decay, and predict a large spectrum of new states of mass $\lesssim 1$ TeV. One might ask if anything can be salvaged in the event that the supersymmetry breaking scale is much higher (e.g. m_{GU} or m_P). A better understanding of the physics behind the "non-renormalization" properties found in supersymmetric theories could perhaps shed some light on this question.

ACKNOWLEDGMENT

The ideas collected here are based on work in collaboration with John Ellis and Bruno Zumino. I am also happy to acknowledge numerous discussions with Murray Gell-Mann on these issues.

This work was supported by the Director, Office of High Energy and Nuclear Physics, Division of High Energy Physics of the U.S. Department of Energy under Contract No. W-7405-ENG-48.

REFERENCES

1. J.C. Pati and A. Salam, Phys. Rev. D8, 1240 (1979).
2. H. Georgi and S.L. Glashow, Phys. Rev. Lett. 32, 438 (1974).
3. For reviews see P. Langacker, SLAC-PUB 2544 (1981), to appear in Physics Reports, and John Ellis, this volume.
4. H. Georgi, H.R. Quinn and S. Weinberg, Phys. Rev. Lett. 33, 451 (1974).
5. A.J. Buras, J. Ellis, M.K. Gaillard and D.V. Nanopoulos, Nucl. Phys. B135, 66 (1978)
6. M.S. Chanowitz, J. Ellis and M.K. Gaillard, Nucl. Phys. B128, 506 (1977).
7. D.V. Nanopoulos and D. Ross, Nucl. Phys. B157, 273 (1979).
8. S. Ferrara, D.Z. Freedman and P. Van Nieuwenhuizen, Phys. Rev. D13, 3214 (1976). S. Deser and B. Zumino, Phys. Lett. 62B, 335 (1976).
9. S.L. Glashow, Nucl.Phys. 22, 579 (1961); S. Weinberg, Phys. Rev. Lett. 19, 1264 (1967); A. Salam, Proceedings 8th Nobel Symposium, Stockholm, ed. N. Swarthholm (Almquist and Wiksells, 1968).
10. D.V. Volkov and V.P. Akulov, Phys. Lett. 46B, 109 (1973); J. Wess and B. Zumino, Nucl. Phys. B70, 39 (1974).
11. M. Gell-Mann, Talk at the 1977 Washington Meeting of the American Physical Society (unpublished).
12. See for example D.Z. Freedman in Recent Results in Gravitation, ed. M. Levy and S. Deser (Plenum Press, N.Y., p. 549, 1978).
13. S. Ferrara, J. Sherk, and B. Zumino, Nucl.Phys. B121, 393 (1977).
14. E. Cremmer and B. Julia, Nucl.Phys. B159, 141 (1979).
15. A. d'Adda, P. Di Vecchia and M. Luscher, Nucl. Phys. B146, 63 (1978); E. Witten, Nucl. Phys. B149, 285 (1979).
16. J. Ellis, M.K. Gailard, L. Maiani and B. Zumino in Unification of Fundamental Particle Interactions, ed. S. Ferrara and P. Van Nieuwenhuizen (Plenum Press, N.Y., p. 69, 1980).
17. An earlier attempt to exploit ref. 14 for GUT phenomenology is P. Cartwright and P. Freund in "Supergravity," Proceedings of the Supergravity Workshop at Stony Brook, Sept. 1979, ed. P. Van Nieuwenhuizen and D.Z. Freedman.
18. A. d'Adda, P. Di Vecchia and M. Luscher, Nucl. Phys. B152, 125 (1979); B. Zumino in Proceedings Einstein Symposium Berlin, Lecture Notes in Physics 100, 114, ed. R. Schrader and R. Seiler (Springer-Verlog, Berlin 1979).
19. M. Veltman, private communication (1980), Similar arguments have also been made by U.T. Cobley, G. t'Hooft, G. Kane, G. Parisi, S. Raby, L. Susskind, and K.G. Wilson.
20. M. Veltman, "The Infrared-Ultraviolet Connection," Univ. of Michigan preprint, to appear in Acta Phys. Polonica. We understand that Kabelschacht is still working on a proof of the theorem.
21. J. Ellis, M.K. Gaillard and B. Zumino, Phys. Lett. 94B, 143 (1980).

22. For a review see B. Zumino, this volume.

23. J. Ellis, M.K. Gaillard and B. Zumino, LAPP preprint in preparation.

24. B. Zumino, Proceedings XX International Conference on High Energy Physics, Madison, Wisconsin, ed. L. Durand and L.G. Pondrom (AIP, New York, 1981) p. 964.

25. S. Ferrara, CERN preprint TH 2957-CERN, Plenary talk given at the 9th International Conference on General Relativity and Gravitation, Jena, G.D.R. (1980).

26. P. Frampton, Phys. Rev. Lett. 46, 881 (1981).

27. M.T. Grisaru and H.J.Schnitzer, Brandeis University preprint, PRINT-81-0572 (1981).

28. H.E. Haber, I. Hinchliffe and E. Rabinovici, Nucl.Phys.B172, 458 (1980); E. Rabinovici, private communication.

29. J.P. Ader, M. Capdeville and H. Navelet, Nuovo Cimento 56A, 315 (1968).

30. J.-P. Derendinger, S. Ferrara, C.A. Savoy, CERN preprint TH-3052 CERN (1981).

31. S. Weinberg and E. Witten, Phys. Lett. 96B, 59 (1980).

32. M.K. Gaillard and B. Zumino, Annecy preprint LAPP-TH-37/CERN-TH-3078, to appear in Nucl. Phys. B.

33. J. Wess and B. Zumino, Phys. Lett. 49B, 52 (1974); J. Iliopoulos and B. Zumino, Phys. Lett. 49B, 52 (1974).

34. T.N. Sherry, Trieste preprint IC/79/105 (1979); S. Dimopoulos and H. Georgi, Harvard preprint HUTP-81/A022 (1981).

COMPOSITE QUARKS AND LEPTONS*

Haim Harari

Weizmann Institute of Science, Rehovot, Israel

ABSTRACT

The possibility that quarks and leptons are composite objects is discussed. The rishon model is studied in detail as an explicit example.

FOREWORD

These notes correspond to a lecture delivered both at the Heisenberg Symposium at the Max Planck Institute in Munich (July 1981) and at the SLAC Summer Institute (August 1981). The lecture relies heavily on the paper "The Rishon Model" by H. Harari and N. Seiberg, Weizmann Institute Preprint WIS-81/38/July-Ph. Most of the work reported here was performed in collaboration with N. Seiberg.

*Work supported by Binational Science Foundation grant no. 49-231

I. Physics Beyond the Standard Model

The "standard model" is based on the local gauge group $SU(3)_C xSU(2)xU(1)$, spontaneously broken to $SU(3)_C xU(1)_{EM}$. It presumably contains three generations of quarks and leptons, twelve gauge bosons (gluons, photon, $W^{\pm}$, Z) and an unknown number of scalar (Higgs) fields. Even if we assume that all future experiments will continue to confirm the predictions of the standard model, we cannot escape the feeling that there must be some new physics beyond it. This conclusion is motivated by the large number of free parameters in the model, by the observed systematics of the spectrum of quarks and leptons, by the unexplained pattern of generations and by the "unnatural" features of a theory containing elementary scalar fields.

Grand unified theories address only the pattern of one generation and actually further complicate some of the other issues. Technicolor schemes introduce composite scalars, but further increase the number of "fundamental" fermions. Supersymmetry ideas are generally believed to be very promising, but no satisfactory concrete models have been proposed so far. Other approaches (extended supergravity, horizontal symmetry, supermultiplets containing several generations) are also interesting but have their problems.

Perhaps the most conservative and uninspired approach would be to suggest that what lies beyond the standard model is another level of compositeness[(1)]. If scalars are composite, one set of problems is solved. If quarks and leptons are composite, their observed spectrum may be explained. The generation pattern may simply represent the excitations of a composite system. The large number of mass and angle parameters should be calculable from the few fundamental constants describing the interactions of the new fundamental fermions.

Needless to say, it is very difficult to construct a self-consistent, realistic composite model of quarks, leptons and scalar particles. Nevertheless, such a program clearly deserves attention, and we would like to study the various problems involved. It is also instructive to pursue a simple model and gain some lessons from its behaviour.

In the next sections, we first discuss some of the general problems associated with the construction of such composite models. We proceed to establish a specific theoretical framework for model building and search for the simplest model which fits into it. We then discuss the general features of the model, and review its successes and difficulties.

II. A Theoretical Framework for a Composite Model

Quarks and leptons are allegedly "pointlike". Evidence from QED, deep inelastic eN and νN scattering, e^+e^- collisions, etc. provides us with limits in the range between 10 GeV and 100 GeV, for the minimal allowed "inverse radius" of quarks and leptons. The excellent agreement between experiment and the QED predictions of the anomalous magnetic moment of the muon, leads to a limit of the order of $r^{-1} \gtrsim 500$ GeV. If quarks and leptons are composite, it is likely that their "compositeness scale" or "inverse radius" is above 500 GeV, possibly many order of magnitude above that value.

In that case, the mass of the composite quark or lepton is significantly smaller than the "compositeness scale" Λ. If the scalar particles are composite and if any of the usual vector bosons are composite, they may also have masses which are small compared with Λ. If we start with a fundamental underlying Lagrangian whose only energy scale is Λ, all of these composite particles are approximately massless on that scale. That can happen "naturally", only if a symmetry principle protects them from acquiring a mass[(2)].

The simplest symmetry principle which forbids fermion mass terms is chiral symmetry. If the basic building blocks of a new underlying theory are massless fermions, the fundamental Lagrangian will possess a certain degree of chiral symmetry. If the chiral symmetry (or a chiral subsymmetry) is not spontaneously broken, some of the composite fermions will be massless.

We therefore look for a model based on fundamental massless fermions bound by a new short range force of characteristic scale $\Lambda > 500$ GeV. The fundamental fermions are presumably "confined" by the new force and are not observable as free particles. The new binding force is analogous to color and is therefore referred to as "hypercolor", based on a nonabelian gauge group which we assume to be $SU(N)_H$.

At energies above Λ_H (the hypercolor scale), the new fundamental fermions are asymptotically free. Below Λ_H, they are confined. They form hypercolor-singlet composites which may or may not possess other quantum numbers such as ordinary color. At "low" energies ($E \ll \Lambda_H$), physics can be described by an effective Lagrangian in which only light hypercolor-singlet particles may appear. These are the composite quarks and leptons as well as composite scalars and the fundamental or composite gauge bosons.

III. Requirements from a Composite Model

(i) A successful composite model must be realistic. It should produce a spectrum of composite quarks and leptons, resembling the observed spectrum, and accounting for the generation pattern.

(ii) The model should be economic in the sense of counting free parameters and basic assumptions.

(iii) The model should be self-consistent in the sense of 't Hooft[2] (and in any other sense...). The calculation of any "flavour anomalies" at the fundamental level and at the composite level should yield the same result. The consistency can be achieved[3] either by complete conservation of the original chiral symmetry (in which case composite massless fermions contribute to the anomaly) or via chiral symmetry breaking (when the composite Goldstone bosons contribute), or, possibly, via chiral breaking followed by conservation of a chiral subsymmetry.

(iv) If Λ_H is significantly larger than the composite masses, the model should be "natural" in the sense that the effective low-energy Lagrangian should be renormalizable, when terms of order Λ_H^{-n} (n-positive) are neglected[4].

(v) The model should yield predictions for experimental quantities such as the proton lifetime, $\sin^2\theta_W$, rare decay rates ($\mu \to e\gamma$, etc.), and other relevant quantities.

We are clearly far from establishing a specific model which has all of these features. However, we shall present our best attempt and outline its important features and its difficulties.

IV. Fundamental or Composite Gauge Bosons?

The gauge bosons of a composite hypercolor theory are the usual gluons, photon, W and Z as well as the new hypergluons of $SU(N)_H$.

The hypergluons are clearly fundamental. They provide the binding of the fundamental fermions, and they must appear in the original Lagrangian. The gluons and photon are massless (like the hypergluons) and they couple to exactly conserved gauge symmetries. The low-energy effective Lagrangian is exactly gauge invariant under $SU(N)_H xSU(3)_C xU(1)_{EM}$ to all orders in $1/\Lambda_H$. It is hard to see how this could happen without a similar invariance of the fundamental Lagrangian. If the fundamental Lagrangian enjoys the same symmetry, the gluons and the photon must appear in it, and they are not composite.

The status of W and Z is different. They are not massless and they correspond to a broken gauge symmetry. There is no compelling reason to assume that all higher dimension terms in the effective Lagrangian are invariant under the full electroweak symmetry (whether it is SU(2)xU(1) or $SU(2)_L xSU(2)_R xU(1)$ does not matter at this stage). It is apriori possible that the electroweak gauge symmetry is a property of the effective Lagrangian when terms of order Λ_H^{-n} are neglected, the fundamental Lagrangian does not respect the same symmetry and the weak gauge bosons may be composite. In that case, the full underlying gauge symmetry of the theory may be $SU(N)_H x SU(3)_C xU(1)_{EM}$, the only fundamental gauge bosons are hypergluons, gluons and the photon and the approximate electroweak symmetry is a consequence of the dynamics of the theory.

V. A Search for a Minimal Model

Starting from the framework of the preceding sections, we now try to find the minimal realistic scheme which obeys the various requirements.

(i) We assume that the full local gauge group is $SU(N)_H xSU(3)_C xU(1)_{EM}$. If the hypergluons, gluons and photon are fundamental gauge particles, nothing less will do. However, we do not allow any additional gauge symmetries.

(ii) We assume that all fundamental massless fermions are in the N-dimensional multiplet of $SU(N)_H$. Antifermions are, of course, in $\bar{N}$-representations.

(iii) In order to form a composite $SU(N)_H$-singlet fermion, we now need at least N fundamental fermions, where N is odd. The smallest odd N is three. Hence, we take the full gauge group to be[5] $SU(3)_H xSU(3)_C xU(1)_{EM}$.

(iv) In order to construct neutral and charged composite fermions, we need at least two fundamental fermions with different electric charges. The fundamental unit of charge is clearly $\frac{1}{3}$e. Hence, the simplest model will involve one fermion with charge $\frac{1}{3}$ (denoted by T) and one neutral fermion (denoted by V), both in hypercolor triplets. We refer to both fermions as "rishons"[6]. The antirishons are hypercolor antitriplets with charges $-\frac{1}{3}$,0.

(v) The only assignments which remain to be determined are the ordinary $SU(3)_C$-colors of T and V. The simplest color representations for each one of them could be 1, 3 or $\bar{3}$. We must form composite quarks and composite leptons belonging, respectively, to (1,3) and (1,1) representations of $SU(3)_H xSU(3)_C$. Hence, we need hypercolor singlets with different color-trialities. Consequently, the two types of rishons must have different colors. Since the definition of 3 and $\bar{3}$ is arbitrary, we are left with only two possibilities: (a) The two types of rishons are in the 1 and 3 color representations, respectively. (b) The two types of rishons are in 3 and $\bar{3}$ color representations. The first possibility leads to a difficulty with Fermi statistics in the construction of leptons. The only remaining possibility is, therefore[5]: T in $(3,3)_{1/3}$ and V in $(3,\bar{3})_0$.

The starting point of our model is, therefore, the following: We postulate a local gauge symmetry based on $SU(3)_H xSU(3)_C xU(1)_{EM}$, with three independent parameters Λ_H, Λ_C, α. There are two types of massless left-handed and right-handed rishons.

The complete list of fundamental particles includes rishons and antirishons, hypergluons, gluons and the photon. All particles are massless. No mass parameters exist. No fundamental scalar particles exist.

The basic model suggested here is clearly extremely economic. The important question is whether, starting from our minimal set of particles, we can construct a realistic self-consistent theory. We try to do that in the following sections. The theory we obtain has many attractive features but it is still far from satisfactory, as many important questions remain open.

VI. The Symmetries of the Fundamental Lagrangian

Our model involves three different local gauge symmetries: hypercolor, color, electromagnetism. They remain exact at all levels of the theory. No other exact symmetry exists.

The underlying Lagrangian possesses, by construction, additional global and discrete symmetries. These symmetries are not necessarily shared by the vacuum, and all of them will be spontaneously broken. We now turn to study these symmetries.

The discrete symmetries of the Lagrangian are very simple. It is clearly invariant under parity and charge conjugation. All fermions appear as both left-handed and right-handed and we have only vector interactions.

The global symmetries are somewhat more complicated. First we note that although we have two types (or "flavors") of rishons, we do not have any global SU(2) symmetry. T and V have identical hypercolor. The two rishons and the two antirishons represent the four possible combinations of hypercolor and color triplets and antitriplets. Each combination appears once and only once.

Superficially, the Lagrangian appears to conserve separately the number of left-handed and right-handed T and V rishons. Hence, we must consider four possible U(1) symmetries for which we may define the following four currents:

$$(J^T_{vector})_\mu = \bar{T}\,\gamma_\mu\,T$$

$$(J^V_{vector})_\mu = \bar{V}\,\gamma_\mu\,V$$

$$(J^T_{axial})_\mu = \bar{T}\,\gamma_\mu\gamma_5 T$$

$$(J^V_{axial})_\mu = \bar{V}\,\gamma_\mu\gamma_5 V$$

The first current J^T_{vector} is actually identical to the electromagnetic current. It therefore represents the local gauge symmetry of $U(1)_{EM}$.

The net number of V-rishons is conserved and the corresponding current J^V_{vector} is a conserved vector current, representing a global $U(1)_V$ symmetry of the Lagrangian.

It will be convenient to define two linear combinations of the net numbers of T-rishons and V-rishons[5]:

$$J^R_\mu = \frac{1}{3}(\bar{T}\gamma_\mu T+\bar{V}\gamma_\mu V)$$

$$J^{B-L}_\mu = \frac{1}{3}(\bar{T}\gamma_\mu T-V\gamma_\mu V)$$

The quantum number R ("rishon number") counts the total number of rishons minus antirishons. The second quantum number counts $[(n_T-n_{\bar{T}})-(n_V-n_{\bar{V}})]$ and turns out to be equal to baryon minus lepton number. We may replace the invariance under $U(1)_{EM}xU(1)_V$ by global invariance under $U(1)_R xU(1)_{B-L}$ where $U(1)_{EM}$ is contained in the product.

The two axial currents J^T_{axial} and J^V_{axial} are not conserved. Their divergences do not vanish because of the well-known anomaly terms:

$$\partial_\mu (J^T_{axial})_\mu = \frac{3g_H^2}{32\pi^2} \varepsilon^{\mu\nu\rho\sigma} (F^a_H)_{\mu\nu} (F^a_H)_{\rho\sigma} + \frac{3g_C^2}{32\pi^2} \varepsilon^{\mu\nu\rho\sigma} (F^a_C)_{\mu\nu} (F^a_C)_{\rho\sigma} +$$

$$+ \frac{e^2}{16\pi^2} \varepsilon^{\mu\nu\rho\sigma} (F_{EM})_{\mu\nu} (F_{EM})_{\rho\sigma} .$$

$$\partial_\mu (J^V_{axial})_\mu = \frac{3g_H^2}{32\pi^2} \varepsilon^{\mu\nu\rho\sigma} (F^a_H)_{\mu\nu} (F^a_H)_{\rho\sigma} + \frac{3g_C^2}{32\pi^2} \varepsilon^{\mu\nu\rho\sigma} (F^a_C)_{\mu\nu} (F^a_C)_{\rho\sigma}$$

Note that the coefficients of the $F_H \tilde{F}_H$ (respectively, $F_C \tilde{F}_C$) terms in both divergences are equal. Again, it is useful to define the sum and differences of the two currents:

$$X_\mu = \bar{T}\gamma_\mu\gamma_5 T + \bar{V}\gamma_\mu\gamma_5 V$$

$$Y_\mu = \bar{T}\gamma_\mu\gamma_5 T = \bar{V}\gamma_\mu\gamma_5 V$$

We immediately observe that while X_μ is not conserved, Y_μ is essentially conserved:

$$\partial_\mu Y_\mu = \frac{e^2}{16\pi^2} \varepsilon^{\mu\nu o\sigma} (F_{EM})_{\mu\nu} (F_{EM})_{\rho\sigma}$$

The current Y_μ is not conserved only as a result of the electromagnetic anomaly term. We can always define a conserved charge $\tilde{Y}$ and our Lagrangian is invariant under the corresponding global axial $U(1)_Y$ symmetry. We therefore have a chiral $U(1)xU(1)$ invariance, which is closely related to our hypercolor and color assignments of the rishons.

The $U(1)_X$ symmetry is broken by color and hypercolor instantons. The Lagrangian does not possess a global $U(1)_X$ symmetry. However, the breaking of the X-charge is always done in "lumps" of 12 units. The instanton term (for color or for hypercolor) must always involve $3T + 3V + 3\bar{T} + 3\bar{V}$ of the same "handedness' for a total $|\Delta X| = 12$. Consequently, a discrete Z_{12} subsymmetry of $U(1)_X$ remains conserved[(7)] (or, in other words, the axial charge X is conserved modulo 12).

The overall symmetry of our Lagrangian is given by $SU(3)_H xSU(3)_C xU(1)_R xU(1)_{B-L} x U(1)_Y$ with parity, charge conjugation and Z_{12} as additional discrete symmetries and a local gauge group $U(1)_{EM}$ contained in $U(1)_R xU(1)_{B-L}$. The global and discrete symmetries need not be respected by the vacuum and may eventually be spontaneously broken.

VII. Composite Quarks and Leptons

At energies well above Λ_H, our theory is asymptotically free. As we decrease the energy, all rishons and hypergluons become confined. Only hypercolor-singlets remain as "free" particles. Composite bosons and fermions are, respectively, constructed from even and odd numbers of fundamental fermions.

What can we say about the masses and radii of such composites? The effective radius of the confining hypercolor force is, presumably, Λ_H^{-1}. Hence, all combinations which actually form composite states are likely to have effective radii of order Λ_H^{-1}. As long as we cannot fully calculate the confining hypercolor forces, we cannot decide whether or not the chiral symmetry is broken, and - in case it is not broken - which composite fermions will emerge massless. At most we may have consistency conditions which should be obeyed by any candidate theory[(2)].

The only exercise we can perform is to check whether the simplest set of composite fermions resembles the observed spectrum of "approximately massless" quarks and leptons. The simplest composite hypercolor-singlet fermions are formed from three rishons or three antirishons. The R and B-L values are determined by the explicit rishon content. For each combination we consider the simplest allowed color representation. The complete list of such three-consistuent composites is given in table 1.

	Hypercolor	1	1
Minimal allowed color	½R / B-L	$\frac{1}{2}$	$-\frac{1}{2}$
1	1	TTT (e^+)	$\bar{V}\bar{V}\bar{V}$ ($\bar{\nu}_e$)
3	$\frac{1}{3}$	TTV (u)	$\bar{V}\bar{V}\bar{V}$ (d)
$\bar{3}$	$-\frac{1}{3}$	TVV ($\bar{d}$)	$\bar{V}\bar{T}\bar{T}$ ($\bar{u}$)
1	- 1	VVV (ν_e)	$\bar{T}\bar{T}\bar{T}$ (e^-)

Table 1: All hypercolor-singlet, three-constituent composite fermions, assuming the minimal color for each configuration.

We immediately notice that the table displays the precise content of one generation of quarks, leptons and their antiparticles. Each B-L value appears with the correct corresponding color and the only allowed combinations are the ones observed in nature.

The relation between the quantization of quark and lepton charges is trivially understood, since all charges come from one fundamental charged particle - the T-rishon.

Another puzzle that is trivially solved here is the famous vanishing sum of all quark and lepton charges in one generation. It is well-known that the renormalizability of the standard model requires the absence of anomalies, leading to the constraint

$$Q(e^-) + Q(\nu) + 3Q(u) + 3Q(d) = 0$$

This constraint is indeed obeyed by a "miraculous" cancellation between quark and lepton charges. In fact, it is the only ingredient of the standard model which tells us that the observed leptons could not have existed without the observed quarks, and vice versa. In our model, the vanishing of the above sum is extremely natural and simple[6]. The total rishon-content of e^-,ν,u,d is $6T+6\bar{T}+6V+6\bar{V}$. Hence, the sum of all electric charges as well as the sum of all B-L values automatically vanish.

Our next concern is to verify that the necessary combinations of rishons obey Fermi statistics. We assume that the effective field of the composite quark or lepton is always constructed from the rishon fields without any derivative coupling. Consider first the leptons and antileptons. Their wave-function contains three identical rishons or antirishons. It is totally antisymmetric under both hypercolor and color. The only remaining degrees of freedom are the Lorentz properties. Each rishon can be described as a Dirac four-spinor transforming according to a $(\frac{1}{2},0)+(0,\frac{1}{2})$ representation of the Lorentz group. The required overall antisymmetry of the wave-function forces the Lorentz part to be totally antisymmetric. The totally antisymmetric product of three $(\frac{1}{2},0)+(0,\frac{1}{2})$ representations is given by one $(\frac{1}{2},0)+(0,\frac{1}{2})$ representation. Consequently, our composite leptons can be "legally" constructed, they must have $J=\frac{1}{2}$, there are no $J=\frac{3}{2}$ leptons and there is only one lepton-state for each set of three rishons[5].

The above discussion tells us that we could not have chosen T and V to be in 3 and 1 representations of $SU(3)_C$ respectively. Such a choice would not have enabled us to form a VVV-lepton with $J=\frac{1}{2}$.

An alternative way to look at the lepton wave-functions would describe a left-handed (or a right-handed) rishon as a two-component spinor. In that case T_L and T_R are different particles. We can, at most, antisymmetrize two T_L-spinors or two T_R's. Hence, we find:

$$e_L^+ = T_R T_R T_L \quad ; \quad e_R^+ = T_L T_L T_R$$

where the two identical rishons always form a Lorentz scalar.

Note that all leptons and antileptons have $Y=\pm 1$. Note also that in each lepton we find a Lorentz-scalar two-rishon set $(T_L T_L)$ or $(T_R T_R)$ or $(V_L V_L)$ etc. This "dirishon" is always antisymmetric in the hypercolor, color and Lorentz degrees of freedom. We will later encounter such "dirishons" again and again in the wave functions of the quarks and the $W^\pm$-bosons.

The analysis of the quark wave-function is more complicated. There are several possible ways of forming a $J=\frac{1}{2}$, color-triplet, TTV composite with a given handedness. Unlike the lepton case, $J=\frac{3}{2}$ quarks cannot be apriori excluded. However, there are general arguments against the existence of massless $J=\frac{3}{2}$ fermions[8]. At this point of our discussion we cannot definitely choose the appropriate wave-function for the u_L-quark. We remark, however, that if we want the quarks to have the same Y-values as the leptons (i.e. $Y=\pm 1$) and to include a "dirishon" system in a color antisymmetric,

Lorentz scalar (as the leptons do), the only allowed solution is given by:

$$u_L = (T_L T_L)V_L \quad ; \quad u_L = (T_R T_R)V_R$$

Our quark and lepton "wavefunctions" are summarized in table 2.

	$SU(2)_L x SU(2)_R x U(1)_{B-L}$	$\frac{1}{2}R$	$\frac{1}{2}Y$	$I_{3L} = \frac{1}{4}(R+Y)$	$I_{3R} = \frac{1}{4}(R-Y)$
$e^+_L = (T_R T_R)T_L$	$(0,\frac{1}{2})_1$	$\frac{1}{2}$	$-\frac{1}{2}$	0	$\frac{1}{2}$
$\bar{\nu}_L = (\bar{V}_R \bar{V}_R)\bar{V}_L$		$-\frac{1}{2}$	$\frac{1}{2}$	0	$-\frac{1}{2}$
$u_L = (T_L T_L)V_L$	$(\frac{1}{2},0)_{1/3}$	$\frac{1}{2}$	$\frac{1}{2}$	$\frac{1}{2}$	0
$\bar{d}_L = (\bar{V}_L \bar{V}_L)\bar{T}_L$		$-\frac{1}{2}$	$-\frac{1}{2}$	$-\frac{1}{2}$	0
$\bar{d}_L = (V_L V_L)T_L$	$(0,\frac{1}{2})_{-1/3}$	$\frac{1}{2}$	$-\frac{1}{2}$	0	$\frac{1}{2}$
$\bar{u}_L = (\bar{T}_L \bar{T}_L)\bar{V}_L$		$-\frac{1}{2}$	$\frac{1}{2}$	0	$-\frac{1}{2}$
$\nu_L = (V_R V_R)V_L$	$(\frac{1}{2},0)_{-1}$	$\frac{1}{2}$	$\frac{1}{2}$	$\frac{1}{2}$	0
$e^-_L = (\bar{T}_R \bar{T}_R)\bar{T}_L$		$-\frac{1}{2}$	$-\frac{1}{2}$	$-\frac{1}{2}$	0

Table 2: Quark and lepton wave-functions and their $SU(2)_L x SU(2)_R x U(1)_{B-L}$ quantum numbers.

VIII. The Weak Interactions

The only fundamental interactions in our model are hypercolor, color and electromagnetism. The W and Z do not exist in the fundamental level of the theory. The short-range weak interaction is not introduced in the underlying Lagrangian and does not operate between single-rishon states.

Consider the forces between two hypercolor-singlet composite objects such as two leptons. At large distances there will be no net hypercolor forces between two $SU(3)_H$-singlets. However, at short distances of the order of Λ_H^{-1} we expect complicated short-range residual forces, reflecting the color and hypercolor interactions between the rishons inside the two composite objects. These forces are analogous to the residual color forces operating between two colorless hadrons. The residual color forces between hadrons are identified with the "strong" or hadronic or nuclear force. We conjecture that the residual forces among hypercolor-singlet composites of rishons be identified with the conventional weak interactions(5). The "tail" of the hadronic forces is determined by the masses of the lightest composite colorless mesons which can be exchanged (i.e. pions, ρ-mesons, etc.). The "tail" of the weak force would now be determined by the masses of the lightest composite hypercolor-singlet bosons which can be exchanged (presumably W and Z).

If our conjecture is correct, the weak interactions should not be considered as one of the fundamental forces of nature. In order to prove such a conjecture, we should be able to derive the observed weak interaction phenomena from our fundamental hypercolor and color forces. This we cannot do at present. However, we are able to study the symmetry properties of the forces among hypercolor-singlet composite fermions.

Let us consider the low-energy effective Lagrangian involving the composite quark and lepton fields. It contains the kinetic terms for the composite fermions as well as the QCD and QED terms describing their interactions with gluons and photons. Consider two composite fermions with the same B-L values, e.g. TTT and $\bar{V}\bar{V}\bar{V}$ (see table 1). Except for the electric charges, all their quantum numbers are identical. It is clear that all the effective Lagrangian terms mentioned above (except for the photon couplings) are invariant under a global SU(2) symmetry, under which the composite fermions form a doublet. If these fermions are massless, the global symmetry is actually $SU(2)_L \times SU(2)_R$.

The quantum number which distinguished between, say, TTT and $\bar{V}\bar{V}\bar{V}$ is the R-number. It is easy to see that(5)

$$\frac{1}{2}R = I_{3L} + I_{3R}$$

We also have a neutral axial charge, given by $I_{3L}-I_{3R}$. We identify it with the axial Y charge of the underlying theory:

$$\frac{1}{2}Y = I_{3L} - I_{3R}$$

We can now easily determine the $SU(2)_L xSU(2)_R$ classification of the quarks and leptons. We find that the left-handed leptons $(\nu_e,e^-)_L$ and the right-handed antileptons $(e^+,\bar{\nu}_e)_R$ are in $(\frac{1}{2},0)$ representation while $(\nu_e,e^-)_R$ and $(e^+,\bar{\nu}_e)_L$ are in $(0,\frac{1}{2})$, and likewise for quarks (see table 2).

What is the overall continuous symmetry of the effective low-energy Lagrangian? At the underlying level we had an $SU(3)_H xSU(3)_C xU(1)_R xU(1)_{B-L} xU(1)_Y$ symmetry. In the effective Lagrangian, the following changes occur:

(i) Only $SU(3)_H$-singlets appear. Hence, the $SU(3)_H$ symmetry is trivial and meaningless.

(ii) A new approximate global $SU(2)_L xSU(2)_R$ appears, containing $U(1)_R$ and $U(1)_Y$ as the two neutral generators.

It is then easy to see that the full approximate continuous symmetry of the effective Lagrangian is $SU(3)_C xSU(2)_L xSU(2)_R xU(1)_{B-L}$. This is precisely the symmetry group of the left-right symmetric extension of the standard model, except that in our case, $SU(2)_L xSU(2)_R xU(1)_{B-L}$ is a global symmetry.

It is remarkable that we are able to find our initial Y-charge inside the global $SU(2)_L xSU(2)_R$ symmetry. It is even more surprising that all quarks and leptons have the correct Y, I_{3L} and I_{3R} values and that the three neutral currents of $SU(2)_L xSU(2)_R x U(1)_{B-L}$ are identical to the three neutral currents of the original Lagrangian corresponding to $U(1)_R xU(1)_{B-L} xU(1)_Y$ or $U(1)_{EM} xU(1)_V xU(1)_Y$.

We summarize: We have conjectured that the weak interactions are residual forces which appear only in the effective Lagrangian. We have shown that, at that level, a new continuous symmetry group exists. It is identical to the left-right symmetric extension of the standard model and the quark and lepton classification is correct. However, the symmetry is global, we do not know whether W and Z bosons exist and we have not yet discussed the symmetry-breaking mechanism.

IX. The Weak Bosons

Our theory involves two levels of Lagrangians. The fundamental Lagrangian is renormalizable. It includes rishons, hypergluons, gluons and the photon. Its continuous symmetry is $SU(3)_H xSU(3)_C xU(1)_R xU(1)_{B-L} xU(1)_Y$, and it describes physics at energies well above Λ_H. In principle, the same Lagrangian should also enable us to calculate low energy phenomena. The low-energy "effective" Lagrangian should be a direct consequence of the basic Lagrangian of the theory.

The "effective" Lagrangian provides us with a phenomenological description of physics well below Λ_H. No hypercolored objects can appear. Hence, the only fields which participate in both levels are the gluons and the photon. To these we must add all hypercolor-singlet composites of mass smaller than Λ_H, such as quarks, leptons and, possibly, additional particles. We have seen that the continuous symmetry of the effective Lagrangian is $SU(3)_C xSU(2)_L xSU(2)_R xU(1)_{B-L}$.

In order to provide us with a realistic theory, the low-energy effective Lagrangian should presumably resemble, as much as possible, the Lagrangian of the standard QCD-electroweak model. Since we already have a global $SU(2)_L xSU(2)_R xU(1)_{B-L}$ symmetry, we would actually expect the effective Lagrangian to approximate the left-right symmetric extension of the standard model(1), based on $SU(2)_L xSU(2)_R xU(1)_{B-L}$, rather than the basic SU(2)xU(1) model.

What are the ingredients of the Lagrangian of the left-right symmetric extension of the standard model?

(i) It includes quarks, leptons, gluons, photon, their kinetic terms and all their interactions (i.e. photon couplings to charged quarks and leptons, gluon-quark couplings and gluon self-couplings).

(ii) It includes six weak bosons ($W_L^{\pm}, W_R^{\pm}, Z_1, Z_2$), their kinetic terms, their self interactions and their interactions with quarks and leptons. Only two coupling constants (g_2 for SU(2) and g_1 for U(1)) describe all of these couplings.

(iii) It includes an unknown number of Higgs fields couple to each of these particles.

Experimentally, we have good evidence for part (i), indirect evidence for (ii) and no evidence for (iii). With parts (i) and (ii) the theory would be renormalizable but all particles remain massless. With parts (i), (ii) and (iii) we regain the full standard renormalizable model.

The effective Lagrangian of the rishon model automatically contains all the terms of part (i). We have already noted that we must also have residual short-range forces operating between hypercolor-singlet fermions. These may be described in terms of effective four-fermion interaction terms. Alternatively, it is possible that composite color-singlet, hypercolor-singlet bosons are formed at masses below Λ_H. Such bosons would then be the carriers of the residual force and would play the role of W's and Z's. We have no specific dynamical reason to expect the creation of such composite bosons, but they may, nevertheless, exist.

If they do, we already know their rishon-content:

$$W_L^+ \equiv (T_L T_L) T_R (V_R V_R) V_L \qquad W_R^+ \equiv (T_R T_R) T_L (V_L V_L) V_R$$

$$W_L^- \equiv (\bar{T}_R \bar{T}_R \bar{T}_L (\bar{V}_L \bar{V}_L) \bar{V}_R \qquad W_R^- \equiv (\bar{T}_L \bar{T}_L) \bar{T}_R (\bar{V}_R \bar{V}_R) \bar{V}_L$$

The parentheses denote a Lorentz-scalar pair of rishons, in an antisymmetric color and hypercolor state. We may also have neutral vector particles with $B-L = R = Y = 0$. They may mix, and we do not apriori know which are the physical eigenstates. Their rishon content can be viewed in terms of $T\bar{T}$ and $V\bar{V}$ combinations, but also in terms of $2r+2\bar{r}$ or $3r+3\bar{r}$ combinations as long as all quantum numbers of all combinations are identical.

What do we know about the couplings and the masses of these "weak bosons"? Since the effective interactions of the weak bosons are residual, their effective couplings should be, in principle, given in terms of complicated functions of the fundamental color and hypercolor coupling constants g_C and g_H (in the same way that the effective ρNN coupling should, in principle, be calculable from QCD).

If the effective Lagrangian is renormalizable when all Λ_H^{-n} terms are removed, the masses and couplings of the W and Z bosons must be those given by the standard electroweak model. We do not know how to prove this or why it happens. This is one of the two main difficulties of the model (the other being the pattern of chiral symmetry breaking). We have no satisfactory answer to this problem at the present time.

Until now we have not discussed the mass terms in the effective Lagrangian. Quark and lepton masses could be due to a small nonvanishing mass of the rishons themselves, or they could be due to dynamical symmetry breaking (or both). The dynamical symmetry breaking could be due to scalar condensates which form as a result of the short-range residual interactions between leptons or quarks. This mechanism could provide us with the correct spectrum of scalar condensates needed in order to reproduce all the phenomenological requirements of an $SU(2)_L xSU(2)_R xU(1)_{B-L}$ gauge theory. We essentially have two types of condensates[9]:

(i) A scalar (or scalars) transforming according to the $(0,1)_{\pm 2}$ representation of $SU(2)_L xSU(2)_R xU(1)_{B-L}$. This condensate is responsible for parity violation, W_R-masses, B-L violation and Majorana masses for neutrinos. All of these phenomena are due only to this scalar field (or fields).

(ii) Scalars transforming like a $(\frac{1}{2},\frac{1}{2})_0$ representation. These contribute to fermion and weak boson masses.

Can we say, at this point, anything about the relationship between g_2 and g_1, the effective weak coupling constants of $SU(2)_L xSU(2)_R$ and $U(1)_{B-L}$, respectively? The underlying Lagrangian is invariant under a global $U(1)_R xU(1)_{B-L}$, which turns out to be a subgroup of $SU(2)_L xSU(2)_R xU(1)_{B-L}$ (since $\frac{1}{2}R = I_{3L}+I_{3R}$). The $U(1)_R$ coupling constant is $g_2/\sqrt{2}$ while the $U(1)_{B-L}$ coupling constant is g_1. The global symmetry group $U(1)_R x$ $U(1)_{B-L}$ is equivalent to the global $U(1)_T xU(1)_V$. However, if $U(1)_R xU(1)_{B-L}$ becomes an approximate <u>local</u> gauge of symmetry, it is equivalent to a <u>local</u> $U(1)_T xU(1)_V$ only if all of these U(1) factors have the same coupling. Hence, the coupling constants of $U(1)_R$ and of $U(1)_{B-L}$ must be equal (or, equivalently, the couplings to T and V must be identical). We then obtain:

$$g_2/\sqrt{2} = g_1$$

The usual definition of $\sin^2\theta_W$ is:

$$\sin^2\theta_W = \frac{g_1^2}{g_2^2+2g_1^2}$$

Hence[5][6]:

$$\sin^2\theta_W = \frac{1}{4}$$

X. Goldstone Bosons and the 't Hooft Condition

We are now in a position to review our various symmetry groups and their symmetry breaking.

It is interesting that, in our model, all exact symmetries are gauge symmetries. The only exact symmetries are, in fact, the three local gauge symmetries of the fundamental theory: hypercolor, color and electromagnetism.

All other symmetries of the original Lagrangian are dynamically broken. The discrete symmetries are parity, charge conjugation and the Z_{12} subgroup of $U(1)_X$. Our scalar condensates break these discrete symmetries, yielding W_R-W_L mass differences, neutrino Majorana masses and Cabibbo mixing which breaks Z_{12}[(7)]. Clearly, no Goldstone bosons are involved in the breaking of discrete symmetries.

The situation concerning the global symmetries is, again, more complicated. Consider the axial $U(1)_Y$ symmetry of the fundamental Lagrangian. At the underlying level, $U(1)_Y$ is a global symmetry group. At the level of the effective Lagrangian we have postulated an approximate gauge invariance under $SU(2)_L x SU(2)_R x U(1)_{B-L}$ which contains $U(1)_Y$ as a subgroup. Hence, $U(1)_Y$ becomes an approximate local gauge symmetry of the effective Lagrangian. The approximate $SU(2)_L x SU(2)_R x U(1)_{B-L}$ gauge symmetry is then dynamically broken by scalar condensates, breaking the $U(1)_Y$ subgroup as well. We know that the dynamical symmetry breaking of a local gauge symmetry (Higgs mechanism) does not require physical massless Goldstone bosons. Thus, in the approximation that $U(1)_Y$ is a local gauge group, we do not expect a Y-breaking Goldstone boson. However, $U(1)_Y$ is not an exact local gauge symmetry of the effective Lagrangian. There will be terms proportional to Λ_H^{-n} which violate the $U(1)_Y$ local gauge invariance but respect the original global $U(1)_Y$ symmetry. Thus we may have a massless Goldstone boson χ, reflecting the breaking of $U(1)_Y$ and coupling to light fermions, with coupling constants of order Λ_H^{-1}. In fact, we must have such a Goldstone boson, since the original theory had an exact global $U(1)_Y$ symmetry and the final outcome shows a breaking of the same $U(1)_Y$.

We can check the consistency of the result in a different way. Define the parameter F_χ as $\langle o|Y_\mu(x)|\chi(q)\rangle = iF_\chi q_\mu e^{-iq\cdot x}$. Clearly, F_χ must be of order Λ_H (what else?). We have the Goldberger-Treiman relation $m_f = g_{\chi ff} F_\chi$ where f is a composite fermion and $g_{\chi ff}$ is the coupling constant of the χ boson to the same fermion. Since in our case $F_\chi \sim \Lambda_H$, $m_f \ll \Lambda_H$, we must indeed have $g_{\chi ff} \sim m_f/\Lambda_H$.

The χ boson is similar to a massless or a low-mass axion with a very small coupling to fermions. It seems that the strongest limit on its existence comes from the process:

$$\gamma + e^- \to \chi + e^-$$

which could frequently take place in the sun (or in stars) and cause energy losses by the emission of the χ-particles.

The obtained limits on Λ_H depend on the χ-mass and on assumptions concerning models of stellar evolution. For instance, for $m(\chi)\lesssim 200$ KeV one gets $\Lambda_H \gtrsim 10^8-10^9$ GeV.

We therefore conclude that, at present, the χ particle does not necessarily pose a major experimental threat to the model. On the other hand, it should be interesting to investigate its theoretical and experimental implications with the hope that they may yield a practical experimental test of the theory.

We should add, at this point, that the appearance of a weakly coupled Goldstone boson is a general feature which must appear whenever a composite model exhibits an underlying global symmetry which becomes a dynamically broken "approximate local gauge symmetry" in the composite level.

In his remarkable paper, 't Hooft has derived[2] necessary consistency conditions relating an underlying theory to its composite companion. In our fundamental Lagrangian, we should consider four anomaly terms, described by the triple products of the currents $Y_\mu Y_\mu Y_\mu$, $Y_\mu J^R_\mu J^R_\mu$, $Y_\mu J^{B-L}_\mu J^{B-L}_\mu$ and $Y_\mu J^R_\mu J^{B-L}_\mu$. Of these, the first three show a vanishing anomaly at the rishon level. Only the last product, $Y_\mu J^R_\mu J^{B-L}_\mu$, has a non-vanishing anomaly.

We must then have composite massless particles which, at the composite level will enable us to obtain the same results. There are two possible ways in which such composite massless particles can account for the anomaly[3]. The first is the existence of massless composite fermions whose contribution exactly provides the needed anomaly. The second way is the existence of a massless composite Goldstone particle which couples directly to the relevant global current and provides the necessary contribution. This is the way in which QCD obeys the anomaly constraint. In our case, a very similar situation exists. Our χ-boson plays the role of the pion in QCD. It couples directly to the current Y_μ and the condition is automatically and trivially obeyed. The composite quarks and leptons are almost, but not exactly, massless. It is debatable whether their contribution to the anomaly should also be included in the consistency equation. However, since all the relevant currents Y, R and B-L, are included in the group $SU(2)_L xSU(2)_R xU(1)_{B-L}$, we know that the total contribution of each generation of quarks and leptons to any anomaly involving these currents, must vanish. Hence, we have no difficulty with the consistency requirements, with or without the contributions of the quarks and the leptons.

XI. Additional Topics

(i) The Generation Structure. The quarks and leptons described above account for one generation. Additional generations are presumably obtained by excitations of the first generation. Orbital and radial excitations clearly have the wrong energy scale. The most likely type of excitation is the addition of rishon-antirishon pairs. We conjecture[7] that the discrete Z_{12} subsymmetry of $U(1)_X$ may provide us with a suitable generation-labelling scheme. We do not repeat here the full discussion of this idea, but the main point is the following: Higher generations can be formed by adding to the first generation a $(T_L\bar{T}_L V_L\bar{V}_L)$ or a $(T_R\bar{T}_R V_R\bar{V}_R)$ Lorentz scalar. All quantum numbers of this Lorentz scalar vanish, except for the X-charge which is changed by ±4 and is conserved modulo 12. In fact, if we insist that each two generations possess different X-values, we can only have three generations. More generally, the number of generations is dictated by the number of colors. The Z_{12}-symmetry is dynamically broken by the same type of scalar condensates which produce the fermion masses, yielding Cabibbo mixing and fermion mass matrices.

(ii) Proton Decay. To lowest order in Λ_H (i.e. $\tau_o \sim \Lambda_H^4$), proton decay is forbidden in our model. It is allowed in higher orders and we obtain[10]:

$$\tau_p \sim \frac{\Lambda_H^8}{M^9}$$

where the M^9 mass factor may involve quark, lepton, proton or W-masses. For different proton decay processes we find that the experimental limit on τ_p yields Λ_H-values around 10^7 GeV.

(iii) Pattern of Chiral Symmetry Breaking. In ordinary QCD with approximately massless u and d quarks the SU(2)xSU(2) "flavor" chiral symmetry is spontaneously broken and an approximately massless Goldstone pion exists. All composite fermions have masses of order Λ_C or more. In contrast, any realistic composite model of quarks and leptons, based on hypercolor dynamics, must exhibit some chiral symmetry which is essentially unbroken, yielding massless composite fermions well below Λ_H. Since the composite fermions do have a small mass, some chiral symmetry breaking must take place, either through small mass terms in the original Lagrangian or through "soft" dynamical symmetry breaking. Regardless of the detailed pattern, one should be able to understand the reason for the difference between the chiral symmetry breaking in QCD and its approximate conservation in our model. We do not have a satisfactory answer to this question. Our model differs from QCD in the interplay between the two color groups, which is crucial for the formation of our set of composite fermions. It is conceivable that this interplay affects the pattern of chiral symmetry breaking, but we are far from a reasonable understanding of this point. This is, probably, the most serious open question of our approach.

(iv) The Limit $g_C \to 0$. It might be interesting to investigate the limit of our model for $g_C \to 0$. We find that our description is very sensitive to such an operation. For $g_C \to 0$, the fundamental Lagrangian acquires a global $SU(6)_L xSU(6)_R xU(1)$ symmetry. When g_C is "turned on", this symmetry breaks into $SU(3)_C xU(1)_R xU(1)_{B-L} xU(1)_Y$. The almost massless quarks and leptons do not form multiplets of SU(6). In fact, there is no SU(6) multiplet whose composite fermion-content remains anomaly-free under the "approximate $SU(2)_L xSU(2)_R xU(1)_{B-L}$ gauge symmetry" of the effective Lagrangian. We must therefore conclude that when $g_C \to 0$, the composite fermion masses move to Λ_H and the chiral symmetry is broken. In a sense this is satisfactory, because the limit $g_C \to 0$ is mathematically completely identical to a six-flavor QCD theory and there we expect the chiral symmetry to break and the fermions to obtain masses of order Λ. In the preceding paragraph we have already conjectured that the difference between QCD and our modol must come in some way from the interplay between the two color groups. If that is the case, sending $g_C \to 0$ would eliminate any such interplay and we should not be surprised if the theory proves to be very sensitive to it. We emphasize that we understand neither the pattern of the chiral breaking nor the $g_C \to 0$ limit. Both are serious difficulties in our model. However, if the interplay of the two colors solves one problem, it automatically solves the other.

XII. Summary

We have attempted here to describe the rishon model as an example of an explicit composite model of quarks and leptons. The model is simple and economic and it reproduces a remarkable number of features of the real world. However - two major obstacles remain and we have no satisfactory explanation for them. The first is the pattern of chiral symmetry breaking which must differ from that of QCD. The second is the problem of light composite weak bosons and the associated approximate gauge symmetry of the low energy effective Lagrangian. We believe that it is worthwhile to continue detailed studies of this and other composite models with the hope of uncovering the physics which lies beyond the standard model.

The ideas discussed in this lecture have been developed in collaboration with Nathan Seiberg. I thank him for many illuminating discussions, suggestions and new ideas.

References

1. For an introductory review see e.g. H. Harari, Proceedings of the 1980 SLAC Summer Institute, A. Mosher (Editor), Stanford, 1981.
2. G. 't Hooft, Lectures at the Cargese Summer School, 1979.
3. Y. Frishman, A. Schwimmer, T. Banks and S. Yankielowicz, Nucl. Phys. B177, 157 (1981).
4. See e.g. M. Veltman, Michigan preprint, 1980.
5. H. Harari and N. Seiberg, Phys. Lett. 98B, 269 (1981).
6. H. Harari, Phys. Lett. 86B, 83 (1979).
7. H. Harari and N. Seiberg, Phys. Lett. 102B, 286 (1981).
8. S. Weinberg and E. Witten, Phys. Lett. 96B, 59 (1980).
9. H. Harari and N. Seiberg, Phys. Lett. 100B, 41 (1981).
10. H. Harari, R.N. Mohapatra and N. Seiberg, CERN TH-3123, July 1981.

RADICAL UNIFICATION

H.P. Dürr
Max-Planck-Institut für Physik und Astrophysik
München (Fed. Rep. Germany)

Abstract

In contrast to the common 'Additive Unification' schemes, like GUTs etc., a 'Radical Unification' scheme - originally suggested by Heisenberg - is adovcated where the fundamental symmetry group and the number of fundamental local fields is required to be minimal. The apparent higher symmetries and the larger number of phenomenological local fields here are effectively generated by the dynamics. In this context a $U(1) \otimes SU(2,loc)$ invariant theory is considered which is based solely on a single 2x2 component isospinor-spinor field ('urfield').

1. Introduction

The unification scheme advocated in the present talk will be called "Radical Unification" in contrast to the more common unification schemes like "Grand Unification Theories (GUTs)", etc. which will be designated as "Additive Unification".

1.1 Additive Unification

Additive unification starts out from the various subtheories for strong, electromagnetic, weak interactions etc. and their corresponding invariance groups. The unification is achieved by combining all these subtheories and then appropriately extending this union to a highly symmetric 'grand' theory such as to achieve invariance under a simple group (in the mathematical sense) and the characterization of the interaction by a single coupling constant. The reduction of the 'grand' symmetry to the direct product of subgroups characterizing the subtheories is enforced by the spontaneous symmetry breaking mechanism steered by an appropriately arranged Higgs-field sector.

Additive Unification hence can be characterized by

- A large fundamental (simple) symmetry group like e.g. SU(5,loc) [1], SO(10,loc) [2], E(6,loc) [3], ... SO(18,loc) [4] ...
- Many fundamental local fields with many components which transform like nontrivial irreducible or reducible representations of this (simple) group. For the SU(5) GUT, for example, the many-component fields representing

  quarks and leptons = $10 + \bar{5}$ (reducible)

  gauge fields = 24

  two types of Higgs fields = $\left\{ \begin{matrix} 24 \\ 5 \end{matrix} \right.$

- A dramatic breakdown of the fundamental symmetry group by a ground state asymmetry (spontaneous symmetry breakdown) due to the condensation of scalar Higgs fields such as to produce the phenomenological symmetry structure. For the SU(5) GUT, for example, the symmetry breakdown occurs in two different steps caused by the condensation of the two Higgs fields

  $SU(5) \xrightarrow{24\text{-plet}} SU(3)_{colour} \otimes SU(2) \otimes U(1)$ 'grand' breaking

  $SU(2) \otimes U(1) \xrightarrow{5\text{-plet}} U(1)_{charge}$ 'flavour' breaking

The phenomenological symmetry structure is supposed to have the form

$$\underset{\text{fermion number}}{U(1)} \otimes \underset{\text{flavour}}{U(1,loc)_Q} \otimes \underset{\text{colour}}{SU(3,loc)} \tag{1.1}$$

or, by removing the asymmetry between weak and electromagnetic interactions, in accordance with the Glashow-Weinberg-Salam-theory (GWS-theory) [5] as

$$U(1) \otimes U(1,loc)_Y \otimes SU(2,loc) \otimes SU(3,loc)_c \tag{1.2}$$

fermion number — flavour — colour

where certain extensions of the group may still be necessary to accommodate characteristic labels of the families (generations).

The simplest GUTs (SU(5) and SO(10), include only one family and do not provide any explanations for the family structure. Higher GUTs [4], on the other hand, which are rich enough to include a certain number of families (e.g. three) require a very large number of Higgs fields to achieve the necessary reduction of symmetry.

The Additive Unification schemes are extremely unsatisfactory from a principal as well as from an aestethic point of view:

- The symmetry groups are huge and their is no indication whatsoever, at present, why a certain large simple group should be more fundamental than any other.
- Because of the large number of fields the theory has many dynamical degrees of freedom which - it is true - appear 'unified' at the outset (at small distances) as long as the symmetry is still intact, but they are ultimately distinguished (at larger distances) by the symmetry breakdown and this as consequence of a completely ad hoc constructed Higgs sector. Therefore, the original gain of unification (one coupling constant) gets completely lost by this arbitrariness.

But what are actually the criteria by which the quality of physical theories should be judged ? This may be to a large extent a matter of taste and adaptation. But let me quote in this context two passages from the autobiography of Albert Einstein [6], the first director of our Institute, which always impressed me very much. He mentions two different view points:

- External confirmation: ... "The theory must not contradict empirical facts. However evident this demand may appear in the first place, its application turns out to be quite delicate. For it is often - perhaps even always - possible to adhere to a general theoretical foundation by securing the adaption of the theory to the facts by means of artificial assumptions ..."
- Inner perfection: "... what may be characterized as the 'naturalness' or the 'logical simplicity' of the premises (the basic concepts and the relations between these). This played an important rôle in the selection and evaluation of theories since time immemorial ... . We prize a theory more highly if, from the logical standpoint, it is not the result of an arbitrary choice among theories which, among themselves, are of equal value and analogously constructed ... "

There are other features of the present approach to particle physics which bother us. In particular, there is what we may call the 'Matryoshka syndrome', i.e. the apparent Russian-doll-in-doll-situation indicated by the monotonous chain: atoms → elementary particles → quarks → subquarks → ... and the continuous quest for the ultimate constituents of matter.

We know that the concept of a constituent and also the distinction between a simple and compound structure, is strictly only meaningful in non-relativistic dynamics [7]. Hence we expect the notion of constituents to fail whenever the relativistic features of the dynamics become relevant. At the level of elementary particles the interaction energies, in fact, do become of the order of the rest mass of the particles and hence we are actually prepared for the constituent description to become very poor if we go beyond. The Matryoshka syndrome had no chance to develop.

If the parton interpretation - or QCD quark theory in its present practical application - however is correct this does not happen because the relativistic features of the dynamics (pair creation etc.) become ineffective on the lower level due to an approximate decoupling at small distances ('free' partons or 'asymptotic freedom' in QCD). The general hope is that this pecularity will continue to happen at a possible subquark level and even further, perhaps with the aid of a sequence of new mass scales with ever increasing mass characterizing the Matryoshka dolls.

To our mind such a situation looks very strange and artificial. One actually suspects that the interpretation of the substructure of elementary particles should be quite different: The approximate decoupling of the relevant modes (quarks etc.) does not indicate a weakening of the basic interaction at small distances but simply indicates that these modes are essentially what one calls 'quasi-particles' in the language of many-body physics. This would also explain why these modes cannot be isolated (confinement).

Anyway, our goal is to avoid, at least in principle, these deficiencies. We, therefore, aim at a rather different unification scheme.

1.2 Radical Unification

In Radical Unification one looks for the common origin or the common root (= radix Gr.) of the effective invariance groups of the various subtheories. This unification corresponds more closely to the German term 'Vereinheitlichung' containing 'Einheit' in the sense of 'oneness' or 'to be one' rather than 'to make one' as in unification. This approach is characterized by the following features:

- A minimally small fundamental symmetry group.
- A minimal number of fundamental local fields with a minimal number of components.

- A mechanism for the effective generation of the phenomenologically observed higher symmetry groups from the fundamental group.
- A mechanism for the effective proliferation of the local fields into the multitude of phenomenological local fields.

The general program in "Radical Unification" distinguished two different levels as depicted in Table I: A fundamental level and an effective phenomenological level which are connected in a complicated, non-perturbative and highly nonlinear fashion.

Fundamental Dynamics

Very small number of fundamental fields
Very small invariance group
Strong nonlinearity (interaction)

Inflation procedure ↓ | ↑ Deflation procedure

Non-perturbative dynamical rearrangement
Effective field proliferation
Effective Generation of higher symmetries
Effective linearization

Effective Dynamics

Large number of effective local fields
Complicated and partly only approximately valid large invariance group
'Weakly' coupled effective fields which allow perturbative treatment

Table I: General Program in Radical Unification

In the following we will at first (section 2) discuss some of the group theoretical properties apparently required to successfully perform the transition from the fundamental to the phenomenological level. Only in the second step (section 3) we will concentrate on the much more complicated question of a fundamental dynamics, and give an example for a dynamical theory which reflects some of the important features. In section 4 a greatly simplified version of this dynamics will be considered and a relationship to the GWS-model of electro-magnetic and weak interactions established. In section 5 I will make some preliminary remarks on the possible interpretation of 'colour' which still poses a problem and finally conclude (section 6) with a few historical comments.

2. Group theoretical considerations

2.1 The fundamental internal symmetry group

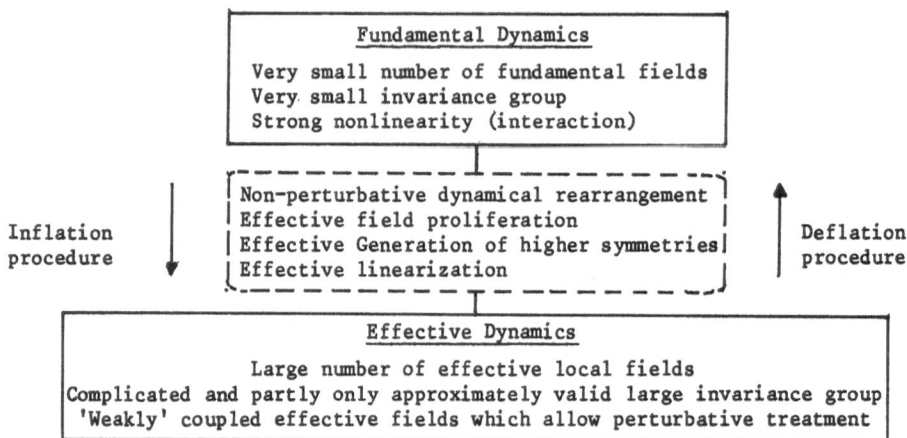

What is the invariance group one has at least to require on the fundamental level to be able to generate, in principle, the rich group structure on the phenomenologicalgical level ? [8] Since there exist symmetry-uprating and symmetry-breaking mechanisms it is difficult to give a clear-cut answer to this question. Hence our policy will be to start with an extremely small invariance group and then investigate how close we can get to describe the phenomenological situation. In fact, we will postulate [9] as basic invariance group the internal symmetry group

$$G = U(1)_F \otimes SU(2,\mathrm{loc})_I \tag{2.1}$$

reflecting a global fermion number F (corresponding to something like B-L) and an isospin I degree of freedom. This is the simplest possibility to introduce a non-abelian gauge property which appears indispensable from general considerations (universal coupling).

2.2 The fundamental set of local fields

The minimal set of local fields which non-trivially represent this basic internal symmetry group (2.1) (besides the Poincaré group as space-time or external symmetry group) is simply one single 2x2 component (anticommuting) isospinor-Weyl-spinor field (a left-handed one, for example) [8,9]

$$\chi(x) = \begin{pmatrix} \cdot \\ \cdot \\ \cdot \end{pmatrix} \mathrel{\hat=} \chi_\alpha(x) \qquad \alpha = 1,2,3,4 \mathrel{\hat=} \alpha'\alpha'' \qquad \begin{matrix} \alpha' = 1,2 \text{ spin} \\ \alpha'' = 1,2 \text{ isospin} \end{matrix} \tag{2.2}$$

The still unmotivated U(1) in (2.1) is automatically provided and is closely related to helicity. In particular, we will not introduce, in addition, an isospin gauge field to secure gauge invariance in the conventional fashion.

This single fermion field - assumed to be quantized by anticommutator conditions - plays the rôle of the fundamental constituent field - we call it 'urfield'*). Conceptually the urfield should be clearly distinguished from a constituent particle which, as mentioned before, has no meaning in the interaction region in a relativistic theory. One should note that the local field concept is different and more general than the particle concept [7, 10] . The local field parametrizes causality and symmetry groups. A particle, on the other hand, has global aspects because it reflects the asymptotic behavior of fields and is connected with stationary solutions of the dynamics involving boundary conditions for their proper definitions.

*) The old germanic prefix 'ur' refers to 'coming out', 'origin', 'source' etc. In the present context it may also be regarded as an abbreviation for 'unificatio radicalis'.

2.3 Symmetry up-rating mechanism

What are the mechanisms which can effectively enlarge the invariance group of the dynamics [11,12] ? From many-body physics one learns that higher symmetries are effectively generated as consequence of an approximate dynamical decoupling of certain subsystems within a system. If the basic dynamics is invariant under a symmetry group G, then a system consisting approximately of n independent subsystems will in the same approximation exhibit the higher invariance group

$$G = \underbrace{G \otimes \ldots \otimes G}_{n \text{ times}} \tag{2.3}$$

where some of the Gs on the r.h.s. may also be subgroups of G.

The best-known example of this type is perhaps the approximate decoupling of spin and orbital motions of the electrons in atoms due to the very small spin-orbit coupling. This leads to an effective bifurcation of the rotation group

$$SU(2) \longrightarrow SU(2)_{spin} \otimes SU(2)_{orbit} \tag{2.4}$$

and the corresponding independent conservation of intrinsic spin $\vec{s}$ and orbital angular momentum $\vec{\ell}$. This independence gets lost for higher energies (strong spin-orbit coupling) such that ultimately only their sum $\vec{j} = \vec{\ell} + \vec{s}$ is conserved.

The 'higher symmetry group' generated in this way by the peculiar structure - we call them structural symmetry in contrast to the basic symmetry - has always the form of iterations of the basic symmetry group or subgroups of these. Atomic, molecular and nuclear physics provide many and more complicated examples for such structural symmetries [12].

2.4 Local field proliferation mechanism

The simplest and most obvious way to effectively generate new local fields is by forming local compounds of the basic field, or formally by considering appropriately defined local products of the local urfield. This mechanism, however, can only be effective if the fundamental interaction is sufficiently strong at small distances to provide sufficient attraction for the binding. Ultraviolet asymptotic free theories do not have this property. Their practical virtue of allowing perturbation theory at small distances turns here to a serious vice by principally barring an effective reduction of the dynamical degrees of freedom.

With the left-handed urfield $\chi(x)$ of (2.2) we can form the following local composites (without derivatives)

2-products	$\vec{\phi}(x) = \tilde{\chi}^{*}\vec{\tau}\chi \hat{=} (\chi\vec{\tau}\chi)$	isovector-scalar
	$\phi_{\mu\nu}(x) = \tilde{\chi}^{*}\sigma_{\mu\nu}\chi \hat{=} (\chi\sigma_{\mu\nu}\chi)$	isoscalar-skewtensor
	$\vec{A}_{\mu}(x) = \chi^{*}\bar{\sigma}_{\mu}\vec{\tau}\chi$	isovector-vector
	$B_{\mu}(x) = \chi^{*}\bar{\sigma}_{\mu}\chi$	isoscalar-vector
(2.5)		
3-products	$\psi(x) = \vec{\tau}\chi\vec{\phi} \hat{=} (\chi\chi\chi)$	isospinor-spinor left-handed
	$\psi(x) = \vec{\tau}\tilde{\chi}\vec{\phi} \hat{=} (\chi^{*}\chi\chi)$	isospinor spinor right-handed
	$\cdots$	
4-products	$h_{\mu\nu}(x) = (\chi^{*}\bar{\sigma}_{\mu}\chi)(\chi^{*}\bar{\sigma}_{\nu}\chi) \hat{=} (\chi^{*}\chi^{*}\chi\chi)$	isoscalar-tensor
...	$\cdots$	
up to 8-products	$\Delta(x) \hat{=} (\chi^{*}\chi^{*}\chi^{*}\chi^{*}\chi\chi\chi\chi)$	isoscalar-scalar

Products higher than 8-products are locally forbidden by the anticommuting property of the urfield. An n-product behaves group theoretically like the anti-(8-n)-product.

This shows the following interesting features:

- The local configurations have at most spin = 2. This is good ! Because we have to represent locally at most 'gravitons'. This limitation is a direct consequence of the internal symmetry requirement (2.1). The requirement of a basic SU(n) would enable locally a maximum spin equal to n. (Derivatives in this context are considered limiting cases of 'non-local' P-wave compounds).
- There is a strong correlation between spin and isospin: Fermions (odd number of urfields) have half-integer isospin, bosons (even number of urfields) have integer isospin. This is bad !

Is there a way to remedy this latter shortcoming ?

2.5 Goldstone field and dressing operators in case of asymmetric ground state

The rigid spin-isospin correlation is, indeed, a very serious difficulty which, as it appears at first, can only be avoided by appropriately extending the basic set of fields, for example by including an additional isosinglet fermion field. A closer investigation, however, reveals that this is not necessary if the ground state is asymmetric under isospin transformations [11,13]. In this case the asymptotic behavior of the theory is dramatically changed and therefore also the character of the effective local fields which interpolate the asymptotic fields.

In case of an asymmetric condensate in the ground state (spontaneous symmetry breakdown), in fact, new local boson fields - the Goldstone fields - can be constructed from the basic fields which correspond to the excitations of the condensate.

In the original Goldstone model [14], for example, the basic field $\phi(x)$ is a nonhermitian scalar field. For the phase-transformation asymmetric ground state $|\Omega\rangle$

characterized by the conditions

(2.6) $$\langle\Omega|\vec{\phi}(x)|\Omega\rangle = \langle\Omega|\vec{\phi}^*(x)|\Omega\rangle = \text{const} \neq 0$$

it is appropriate to decompose it into its 'radial' and 'azimuthal' part

(2.7) $$\phi(x) = \rho(x)\exp[i\Theta(x)]$$

The azimuthal part plays the rôle of the Goldstone field which under condition (2.6) allows the formal expansion in terms of the basic field

(2.8) $$\Theta(x) = \tan^{-1}\left[i\frac{\phi-\phi^*}{\phi+\phi^*}\right] = i\frac{\phi-\phi^*}{\phi+\phi^*} + \cdots = \frac{\mathrm{Im}\,\phi}{\langle\phi\rangle} + \cdots$$

This local construction of the Goldstone field seems meaningful if $\langle\phi\rangle = \langle\phi^*\rangle$ represents the dominating part of the field operator and hence should fail for distances smaller than m_{Higgs}^{-1}, with m_{Higgs} the mass of the quantized part of the Higgs field.

The Goldstone fields have the property that they transform essentially <u>additive</u> under the broken symmetry transformations,

(2.9) $$\begin{cases} \phi(x) \mapsto \exp[i\alpha]\,\phi(x) \\ \Theta(x) \mapsto \Theta(x) + \alpha \end{cases}$$

In case of a fermion pair condensate (like in superconductivity) the Goldstone field can be similarly expressed in terms of local products of the fermion fields [13].

In the <u>GWS-model</u> [5], where an isodoublet scalar Higgs field $\varphi_\alpha(x) \neq \varphi^*_\alpha(x)$ is introduced, the asymmetry condition

(2.10) $$\langle\Omega|\varphi_\alpha(x)|\Omega\rangle = \langle\Omega|\varphi^*_\alpha(x)|\Omega\rangle = \lambda\binom{0}{1} \neq 0$$

leads to a double breaking of the symmetry according to

(2.4) $$U(1)_Y \otimes SU(2)_T \underset{\substack{T_1,T_2\\ \text{breakdown}}}{\Longrightarrow} U(1)_Y \otimes U(1)_{T_3} \underset{\substack{Y-T_3\\ \text{breakdown}}}{\Longrightarrow} U(1)_{Q=Y+T_3}$$

Similarly to (2.8) [13] this gives rise to 2+1 Goldstone fields as functionals of φ_α and φ^*_α

(2.12) $$\begin{cases} \Theta_\pm(x) = \Theta(x)\exp[\mp i\varphi(x)] = \Theta_\pm(\varphi_\alpha, \varphi^*_\alpha) \\ \sigma(x) = \sigma(\varphi_\alpha, \varphi^*_\alpha) \end{cases}$$

The three Goldstone fields $\theta(x)$, $\varphi(x)$, $\sigma(x)$ are closely related to the three Euler angles.

With the Goldstone fields local 'dressing operators' can now be constructed [13,15] with a mixed transformation character: They partially transform like vectors under the original group and partially like vectors under the invariance group of the ground state (stability group). For the Goldstone model it is simply the operator

$$\mathcal{U}(\Theta(x)) = \exp[i\,\Theta(x)] \mapsto \exp[i\alpha]\,\mathcal{U}(\Theta(x)) \tag{2.13}$$

and for the GWS-model the 2-dimensional rotation matrices with the Euler angles $\theta(x)$, $\varphi(x)$, $\sigma(x)$

$$\begin{cases} S(\theta_\pm(x), \sigma(x)) = S(\theta(x), \varphi(x), \sigma(x)) \\ \qquad \hat{=}\ S_\alpha^{\underline{\alpha}}(x) = \left(\begin{smallmatrix}\cdot & \cdot\\ \cdot & \cdot\end{smallmatrix}\right) \\ \qquad = S_0(\theta_\pm(x))\,\mathcal{U}(\sigma(x)) \end{cases} \tag{2.14}$$

These dressing operators like the Goldstone fields should be regarded as quasilocal operators which loose their meaning for distances $\ll m_{Higgs}^{-1}$.

Because of the mixed transformation property these dressing operators have the property of transmutators [13] in the sense, that they transmute the 'live' U(1) ⊗ SU(2) index 'α' into a 'frozen' index '$\underline{\alpha}$' of the U(1) invariance group of the ground state. They have some similarity to the 'Killing vector fields' in differential geometry.

With these dressing operators 'frozen' fields [13,16] can be constructed (sometimes also referred to as 'bleaching'). In the GWS-model, for example, the isodoublet fermion field $\chi_\alpha(x)$ can be transmuted into two singlets $\chi_{\underline{\alpha}}(x)$ ($\underline{\alpha}$ = 1,2) according to

$$\chi_{\underline{\alpha}}(x) = S^{*}(x)_{\underline{\alpha}}{}^{\alpha}\,\chi_\alpha(x) \tag{2.15}$$

Since the isospin asymmetric ground state and hence the Goldstone mode can equally well arise from a condensation of a pair of isodoublet fermions, the particular construction (2.15) now allows to explicitly build isosinglet fermions $\chi_{\underline{\alpha}}(\chi_\alpha, \chi_\alpha^*)$ from isodoublet fermions. Hence integer isospin fermions and half-integer isospin bosons can now be constructed from the urfield [13].

2.6 Hypercharge

The latter construction, in fact, admits an interesting possibility for the interpretation of hypercharge [13,15].

Hypercharge Y, defined as the medium charge of a multiplet

$$Q = Y + T_3 \tag{2.16}$$

has a strange hybrid character in the sense that on the one hand it is connected with local transformations and on the other hand with fermion-number type global transformations, as e.g. in the Gell-Mann-Nishijima charge formula. It has halfintegral values for leptons and 1/6-values for quarks.

In the framework described above hypercharge can now be simply considered as 'stripped', 'shielded', 'frozen' or 'bleached' isospin effected by the $\mathfrak{S}(\theta \pm (x))$ transmutator defined in (2.14)

(2.17) $$\mathfrak{S}\,\alpha^{\alpha}(x) = \begin{bmatrix} \cos\theta & -\sin\theta e^{-i\varphi} \\ \sin\theta e^{i\varphi} & \cos\theta \end{bmatrix}$$

which does not affect the I_3-transformations, i.e.

(2.18) $$\vec{I} \xrightarrow[\text{stripping}]{I_1, I_2} \vec{I}_{\text{stripped}} \overset{!}{=} Y$$

Hypercharge reflects so-to-say the left-over I_3-property after I_1, I_2 bleaching. As a consequence charge is simply identical to isospin-charge

(2.19) $$Q = I_3 = (I_3)_{\substack{\text{shielding} \\ \text{Goldstone cloud}}} + (I_3)_{\text{field}}$$

$$= Y + T_3$$

Therefore the local hypercharge group $U(1,loc)_Y$ in the GWS-model may be redundant [15].

The second dressing operator $u(\sigma(x))$ in (2.14) has the property of transmuting fermion number into hypercharge

(2.20) $$F \longleftrightarrow Y$$

and hence offers an explanation for the hybrid character of hypercharge.

2.7 Vector gauge fields

The invariance under local symmetry transformations like SU(2,loc) usually requires the introduction of a compensating vector gauge field $A_\mu(x)$ which transforms inhomogeneously (like a connection) under the group

(2.21) $$\begin{cases} \chi(x) \longmapsto \exp\left[-i\frac{\vec{\tau}}{2}\cdot\vec{\beta}(x)\right]\chi(x) \\ \vec{A}_\mu(x) \overset{\text{inf}}{\longmapsto} \left[1 + \vec{\beta}(x)\times\right]\vec{A}_\mu(x) + \partial_\mu\vec{\beta}(x) \end{cases}$$

Local products of $\chi(x)$ on the other hand always transform (homogeneously like tensors. Hence it appears that gauge fields have to be introduced independently as basic fields. This, however, is not necessarily the case.

The anticommutator of the urfield $\chi(x)$ with $\chi^*(x)$ is assumed to be nontrivial to secure quantization in contrast to the anticommutator of χ with χ, or χ^* with χ^*. The 2-point-function $\chi(x+\xi/2)\,\chi^*(x-\xi/2)$, therefore, must have a c-number-part singular for $\xi \to 0$. As a consequence local products of operators χ^* with χ have to be appropriately defined as 'finite parts' by extracting the divergent singular parts. Because of this 'regularization' one can show [17] that

(2.22) $$\vec{A}_\mu(x) \equiv :\chi^+ \bar{\sigma}_\mu \vec{\tau} \chi:(x)$$

($\bar{\sigma}_\mu = I, -\vec{\sigma}$ Pauli matrices) transforms exactly like a gauge field <u>if</u> one requires the urfield to have the subcanonical (inverse length) dimension (behavior under dilatation)

(2.23) $$\dim\ \chi(x) = \frac{1}{2}$$

in contrast to the dimension 3/2 of a canonical spinor field. This condition can only be realized in a state space with indefinite metric*).

The dimension of the gauge field will be 'canonical'

(2.24) $$\dim\ A_\mu(x) = 1$$

as it should.

The scalar Higgs field constructed as an S=0, I=1 urfield pair

(2.25) $$\vec{\phi}(x) = \tilde{\chi}^+(x)\vec{\tau}\chi(x) = \chi^T(x)\,c_\sigma^* c_\tau^*\,\vec{\tau}\chi(x) = :\tilde{\chi}^*\vec{\tau}\chi:(x)$$

does not require any subtractions and hence transforms homogeneously. It also acquires canonical dimension

(2.26) $$\dim \phi(x) = 1$$

2.8 Summary

The main results of this section are summarized in Table II. From the single urfield $\chi(x)$ and its hermitian conjugate field we can construct a whole set of local

---

*) There are more sophisticated ways of defining the 'finite part' of an operator product where the dimension of the product is not simply given in the naive way by the sum of the dimensions of the factors. This allows to push this 'difficulty' into a different corner without really avoiding it.

fields with various transformation properties by forming appropriate local products. In case of an asymmetric ground state this set of local fields can be effectively extended to a larger set with the aid of quasi-local dressing operators (transmutators) which allow partial bleaching of the 'broken' degrees of freedom. This procedure leads to a rather rich assortment of effective local fields which should be compared with the local field operators of the conventional theories.

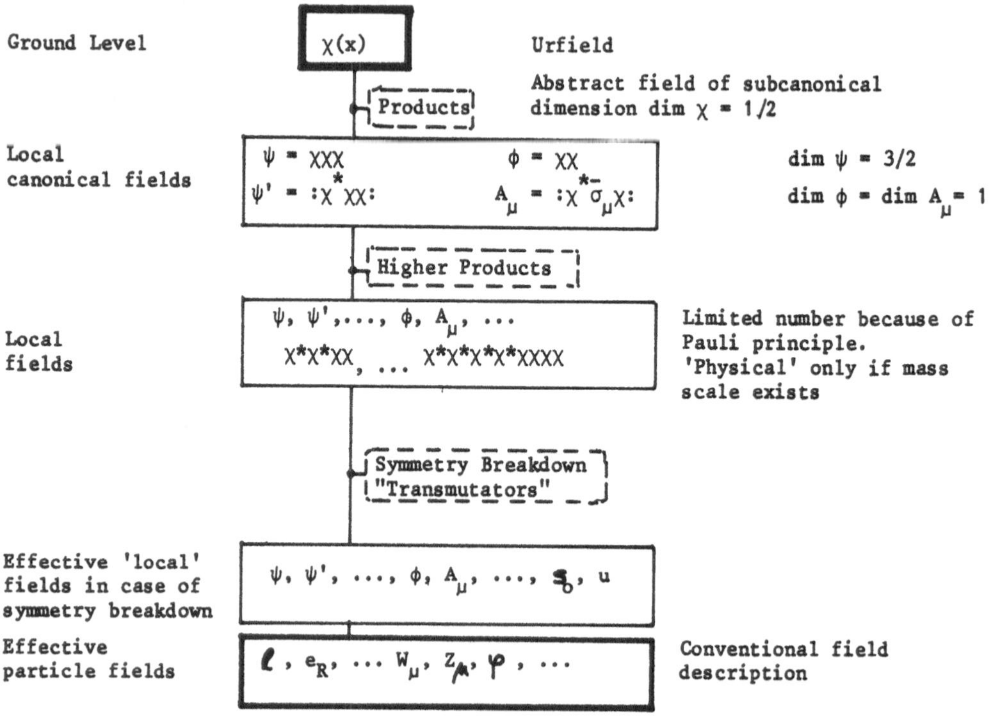

Table II: Construction of effective local fields from the urfield

We have, for example, the explicit constructions reflecting the correct transformation properties (for details see [9,18,19,20])

$$
(2.27)\quad
\begin{cases}
\text{left-handed leptons} & \ell = \begin{pmatrix}\nu\\ e_L\end{pmatrix} \sim u\vec{\tau}\chi\vec{\phi} \sim u\chi\chi\chi \\
\text{right-handed leptons} & e_R \sim u^3(\xi_0^*\vec{\tau}\hat{\chi})\vec{\phi} \sim u^3\xi_0^*\chi^*\chi\chi \\
\text{left-handed quarks} & q = \begin{pmatrix}u_L\\ d_L\end{pmatrix} \sim u^{*1/3}\vec{\tau}\chi\vec{\phi}^* \sim u^{*\frac{1}{3}}\chi\chi^*\chi^* \\
\text{right-handed quarks} & \begin{cases} q_1 = u_R \sim u^{*7/3}\tilde{\xi}_0^*\vec{\tau}\chi\vec{\phi} \sim u^{*7/3}\xi_0\chi^*\chi\chi \\ q_2 = d_R \sim u^{5/3}\tilde{\xi}_0^*\vec{\tau}\chi\vec{\phi}^* \sim u^{5/3}\xi_0\chi^*\chi^*\chi^* \end{cases} \\
\ldots
\end{cases}
$$

$$
(2.28)\begin{cases}
\text{GWS-Higgs field} & \varphi \sim u^{*2}\vec{\tau}\tilde{\underset{\circ}{s}}\vec{\phi} \sim u^{*2}\underset{\circ}{s}XX \\
\text{GWS-gauge fields} & \begin{cases} \vec{W}_\mu = \vec{A}_\mu + (\vec{A}_\mu - \vec{\underset{\circ}{B}}_\mu)^\perp \\ Z_\mu = -\vec{S}\cdot\vec{A}_\mu + 2U_\mu + \underset{\circ}{B}_\mu \end{cases} \\
\vec{S} = \underset{\circ}{s}^*\vec{\tau}\underset{\circ}{s}\ , \ \vec{\underset{\circ}{B}}_\mu = i\underset{\circ}{s}^*\vec{\tau}\overleftrightarrow{\partial}_\mu\underset{\circ}{s}\ , \ \underset{\circ}{B}_\mu = \vec{S}\cdot\vec{\underset{\circ}{B}}_\mu\ , \ \vec{\underset{\circ}{B}}_\mu^\perp = \vec{\underset{\circ}{B}}_\mu - \vec{S}\underset{\circ}{B}_\mu & \\
\dots &
\end{cases}
$$

The operators indicated here for the quarks are still unsatisfactory because they reflect only their flavour properties but not their colour properties. This is a very serious defect for which no simple remedy can be offered, at present. There are some indications, however, how this problem may be resolved. We will shortly return to this point at the end (section 5).

It should also be emphasized that our considerations up to now were purely kinematical. We have simply looked into the question: Given a single urfield operator $\chi(x)$, what local or quasilocal configurations can possibly be constructed from it ? Or turning the problem around: Given the phenomenologically 'established' local fields, how can they be interpreted as particular compounds of a single urfield ? We may call the latter procedure: Deflation of the physical fields. The answer is: Up to the colour property all known physical fields can, in principle, result from a single urfield (2.2).

This first step of our radical unification program has to be clearly distinguished from the much tougher and extremely more difficult problem: What is the basic dynamics of the urfield which will enable these particular configurations to establish themselves as 'bound states', or more precisely, as particles with the correct masses and the correct coupling strengths in their mutual effective interactions ?

Clearly such a dynamics has to be strongly nonlinear in order to produce sufficiently strong interactions at small distances. In the next section we will offer some speculations on such a fundamental dynamics.

3. Urfield Dynamics

To make a good guess for the dynamics of the underlying urfield inducing all the known phenomenological features appears prohibitively difficult. The task, indeed, seems ultimately more complicated than, for example, extracting the Schrödinger equation from the spectrum of the uranium atom or the DNS-molecule, because we cannot hope for an approximate decoupling of the fundamental degrees of freedom in this case. In this desperate situation the only chance we have to make a good hit is the hope that the fundamental dynamics is clearly distinguished from other possibilities by some 'naturalness' or 'logical simplicity', as Einstein demanded. In the present context this would mean that the urfield dynamics exhibits maximum symmetry in some sense and does not involve any arbitrary dimensionless numerical constant - if this can be achieved, at all.

3.1 Urfield Lagrangian

It is interesting to note that for the 2x2 component nonhermitian isospinor-spinor urfield (2.2) there does, indeed, exist an extremely simple and unique dynamics [8,9,21].

Because of the anticommutativity property of the urfield there exist the two nonhermitian highly symmetrical (SL $(4,\mathbb{C})$ invariant) products formed with the alternating Levi-Civita symbol

$$(3.1)\qquad \begin{cases} X(x) = \varepsilon^{\alpha\beta\gamma\delta}\chi_\alpha(x)\chi_\beta(x)\chi_\gamma(x)\chi_\delta(x) \\ X^*(x) = \varepsilon^{\dot\alpha\dot\beta\dot\gamma\dot\delta}\chi^*_{\dot\alpha}(x)\chi^*_{\dot\beta}(x)\chi^*_{\dot\gamma}(x)\chi^*_{\dot\delta}(x) \end{cases}$$

reflecting the fact that, due to the Pauli principle, each space-time point x can be maximally occupied by four χs and by four χ^*s, if χ and χ^* are 4-component fields. For the local Lagrangian hence the following hermitian product is uniquely suggested

$$(3.2)\qquad \begin{cases} \mathcal{L}(x) = \,:X(x)X^*(x): \\ \qquad = \varepsilon^{\alpha\beta\gamma\delta}\varepsilon^{\dot\alpha\dot\beta\dot\gamma\dot\delta}:\chi_\alpha\chi_\beta\chi_\gamma\chi_\delta\chi^*_{\dot\alpha}\chi^*_{\dot\beta}\chi^*_{\dot\gamma}\chi^*_{\dot\delta}:(x) \\ \qquad = 4!:\det[\chi_\alpha\chi^*_{\dot\beta}]:(x) \end{cases}$$

i.e. an expression which has essentially the structure of the 4x4 determinant constructed from χ and χ^*. Since the anticommutator

$$(3.3)\qquad \lim_{\xi\to 0}\{\chi(x+\xi/2),\chi^*(x-\xi/2)\} \neq 0$$

has to be nonzero to secure quantization, local products of χ and χ^* - in contrast to χ-products or χ^*-products - are necessarily singular and therefore require for their proper definition an appropriate 'finite part' prescription indicated by : : .

In order to establish dilatation invariance for the action

$$(3.4)\qquad \dim \int d^4x\,\mathcal{L}(x) = 0 \qquad\qquad \text{or } \dim \mathcal{L}(x) = 4$$

the urfield has - by naive argument - carry just the subcanonical dimension mentioned earlier

$$(3.5)\qquad \dim\chi = \dim\chi^* = \frac{1}{2}$$

This requires that the 2-point function of the urfield has a leading c-number part (propagator) of dimension 1:

(3.6)
$$\chi(x+\xi/2)\chi^*(x-\xi/2) = \underbrace{i\frac{\sigma\cdot\xi}{\xi^2}\tau^0}_{\text{'propagator'}} + \underbrace{:\chi(x+\xi/2)\chi^*(x-\xi/2):}_{\text{'finite part'}}$$

The subcanonical propagator corresponds essentially to a 'dipole ghost' propagator [22]

(3.7)
$$\underbrace{\chi\chi^*} \sim \int d^4p\, \frac{\sigma\cdot p}{(p^2)^2} \exp[-ip\cdot\xi]$$

The decomposition (3.6) implicitly contains a definition of the finite part of the $\chi\chi^*$-product

(3.8)
$$:\chi\chi^*:(x) = \lim_{\xi\to 0}\left(\sigma\cdot\frac{\partial}{\partial\xi}\right)(\bar{\sigma}\cdot\xi)\chi(x+\xi/2)\chi^*(x-\xi/2)$$

The $\chi\chi$- and $\chi^*\chi^*$-products are finite without subtractions.

(3.9)
$$:\chi\chi:(x) = \chi(x)\chi(x) \quad , \quad :\chi^*\chi^*:(x) = \chi^*(x)\chi^*(x)$$

Definitions (3.8, 3.9) can be used to write the Lagrangian (3.2) in the alternate form

(3.10)
$$\mathcal{L}(x) = \lim_{\xi\to 0} \partial_\xi^2\partial_\xi^2(\xi^2)^2 \det[\chi(x+\xi/2)\chi^*(x-\xi/2)]$$

Since differentiations with regard to the split vector ξ produce 'interior' derivatives $\chi\overleftrightarrow{\partial}\chi^*$ one realizes that in consequence of the finite part prescription the Lagrangian obtains effectively the conventional form involving also derivatives of the fields (in the present case, in fact, up to the third derivative).

3.2 Symmetry of the urfield Lagrangian

The Lagrangian (3.2) or (3.10) has an extremely high symmetry. It is not only invariant under Poincaré transformations and dilatations (due to (3.5)) but, in fact, under the full 15-parameter conformal group [21]. In addition - due to the determinant structure $\det[\chi(x_+)\chi(x_-)]$ as depicted in (3.10) - it appears to be invariant under the huge gauge-type group $U(1) \otimes SL(4,\mathbb{C}, \text{loc})$. However, because of the limiting procedure, it is reduced to the group

(3.11)
$$G = U(1)_F \otimes SU(2,\text{loc})_I$$

It is interesting to note:

- Dilatation invariance and uniqueness of the propagator in space-time (no cuts) establishes, via the Pauli principle, a relationship between number of flavours n and space-time dimension D as

$$n = \frac{1}{2} D = 2 \tag{3.12}$$

singling out the isospin-doubling of the urfield*).

- An invariance under a non-abelian gauge symmetry occurs <u>without</u> the explicit appearance of a corresponding vector gauge field.

The latter may be rather surprising. One has to realize, however, that vector gauge fields occur in connection with derivatives which are only implicit in the form (3.10) of the Lagrangian. To make a comparison with conventional theories we hence should look at a derivative explicit form of the Lagrangian.

3.3 Derivative form of the urfield Lagrangian

The derivative form of the Lagrangian (3.10) is obtained by explicitly carrying out the ξ-limit [21]. A differentiation with regard to the split-vector ξ will essentially produce an 'interior' derivative. The Lagrangian obtains roughly the structure

$$(3.13)\quad \begin{cases} \mathcal{L}(x) \sim -:\chi^*\left(\frac{i}{2}\right)^3 \bar{\sigma}\cdot\overleftrightarrow{D}\,\overleftrightarrow{D}^2\chi: + \frac{4}{9}:\psi^*\frac{i}{2}\bar{\sigma}\cdot\overleftrightarrow{D}\psi: - \\ \qquad - 3|D_\mu\vec{\phi}|^2 - 3|D_\mu\phi^{\kappa\lambda}|^2 - \frac{1}{48}(\vec{\phi}^2 - \phi^2_{\kappa\lambda})(\vec{\phi}^{*2} - \phi^{*2}_{\kappa\lambda}) \\ \quad = \frac{4}{9}:\psi^*\frac{i}{2}\bar{\sigma}\cdot\overleftrightarrow{D}\psi: - 3|D_\mu\vec{\phi}|^2 - \frac{1}{48}\vec{\phi}^2\vec{\phi}^{*2} + \dots \end{cases}$$

involving the 'covariant derivative'

$$(3.14)\quad \begin{cases} D_\mu = \partial_\mu + i A_\mu \\ A_\mu = \frac{\vec{\tau}}{2}\cdot\vec{A}_\mu - \frac{1}{8}\sigma_{\kappa\lambda}A^{\kappa\lambda}_\mu - \frac{1}{8}\sigma_{\kappa\lambda}\vec{\tau}\cdot\vec{A}^{\kappa\lambda}_\mu \end{cases}$$

Here the $\vec{\phi}, \phi_{\kappa\lambda}, \vec{A}_\mu$ are essentially the bilinear forms and ψ the trilinear form of the urfield as given before in (2.5) and (2.22). One easily checks that the SU(2,loc) gauge invariance here is exactly established in the way described earlier by the occurence of the inhomogeneously transforming vector gauge field $\vec{A}_\mu$ (2.22).

The form (3.13) is only one way of writing the Lagrangian. There are other forms depending what kind of effective local fields we introduce. In particular we may also use the canonical spinor field $\psi' \sim \chi^*\chi\chi$ besides $\psi \sim \chi\chi\chi$. The appropriate choice of fields should, of course, be determined by the dynamics itself.

---

*) An SU(n) internal symmetry, in general, would require $\mathbf{X} \sim \chi^{2n}$ and therefore dim $\chi = 1/n$ which in the 2-point function required the denominator $(\xi^2)^{1/n+1/2}$ and hence produces cuts except for n = 2.

The Lagrangian (3.13) has some similarity to the GWS-Lagrangian, but there are some distinct differences, e.g.

- There is a third derivative term of the urfield. This establishes the self-consistency of the dim $\chi = \frac{1}{2}$ postulate.
- The Higgs fields occur in different forms, in particular as $\vec{\phi} \sim (\chi\vec{\tau}\chi)$ but also as $\phi_{\mu\nu} \sim (\chi\sigma_{\mu\nu}\chi)$ rather than in the isodoublet-scalar form.
- Besides $\vec{A}_\mu$ there are additional 'gauge fields' $A_\mu^{\kappa\lambda} \sim :\chi^* \sigma_\mu \sigma^{\kappa\lambda}\chi:$ and $\vec{A}_\mu^{\kappa\lambda} \sim :\chi^* \sigma_\mu \sigma^{\kappa\lambda}\vec{\tau}\chi:$ reflecting a gauge structure in the 4-dimensional spin-isospin space (connected with SL(4,$\mathbb{C}$, loc) invariance of the naive Lagrangian). The invariance group, however, is only SU(2,loc)$^{*)}$.
- The kinetic terms of the Higgs fields have opposite sign in comparison with the canonical ansatz. This may have severe consequences regarding the existence of the W- and Z-vector-bosons.
- There are no kinetic terms for the gauge fields $A_{\mu\nu}A^{\mu\nu}$ etc. One can easily convince oneself that such kinetic terms only occur after an iteration of the Lagrangian because they necessarily require two independent split vectors (plaquettes). It is not clear, however, whether such terms are really required, except, of course, for the photons. The absence of such terms would mean that no W- and Z-vector-boson will be observed (in this context see also Saller [19]).

The differential form (3.13) of our simple Lagrangian (3.10) is rather complicated. To extract some dynamical consequences we therefore will limit ourselves here to consider a modified and simplified version of this Lagrangian.

4. Model Lagrangian

We investigate the following U(1) ⊗ SU(2,loc) invariant model Lagrangian [22]

$$
(4.1) \quad \begin{cases}
\mathcal{L}(x) = \psi^* \sigma^\mu \left(\frac{i}{2}\overleftrightarrow{\partial}_\mu - \frac{\vec{\tau}}{2}\cdot\vec{A}_\mu\right)\psi - \frac{1}{4g_0^2}\vec{A}_{\mu\nu}^2 + \\
\qquad + \frac{1}{2}\left|(\partial_\mu - \vec{A}_\mu \times)\vec{\phi}\right|^2 - V(\vec{\phi}) - \\
\qquad - G_2\left[\vec{\phi}^*\cdot(\tilde{\psi}^*\vec{\tau}\psi) + (\psi^*\vec{\tau}\tilde{\psi})\cdot\vec{\phi}\right] \\
V(\vec{\phi}) = \frac{1}{4}\bar{h}\left[|\vec{\phi}|^2 - M^2\right]^2 + \frac{1}{4}(h - \bar{h})\vec{\phi}^2\vec{\phi}^{*2} \\
\vec{A}_{\mu\nu} = \partial_\nu \vec{A}_\mu - \partial_\mu \vec{A}_\nu + \vec{A}_\mu \times \vec{A}_\nu
\end{cases}
$$

---

*) There is an extension of the gauge invariance group which includes the local Lorentz group if one explicitly introduces the vierbein as independent field. Here then the $A_\mu^{\kappa\lambda}$ are really 'connections' and relate to torsion.

This is a GWS-type Lagrangian which, however, simulates some of the most important features of our effective urfield Lagrangian, in particular it contains

- Only an SU(2,loc) gauge group with corresponding isovector gauge field $\vec{A}_\mu$ (no hypercharge)
- Only left-handed leptons
- An isovector Higgs field which allows a bifermionic $(\psi\psi)$-interpretation with selfcoupling parameters $\bar{h}$ and h. (The urfield case corresponds to the special case $\bar{h} = 0$).

Under certain conditions, which are given elsewhere [23], one can demonstrate that this Lagrangian does contain all the salient features of the GWS-model with the specifications

- Charge is only isospin-charge: hypercharge is frozen (or bleached) isospin.
- The local fields proliferate (approximately and presumably for distances $\gg m_{Higgs}^{-1}$) in such a way as to provide, in addition, a right-handed isosinglet electron as 'frozen' anti-lepton and a hypercharge vector gauge field Z_μ essentially as 'frozen' $\vec{A}_\mu$. The coexistence of local field, in the non-frozen <u>and</u> frozen form ('hard' dressing [24]) is governed by Goldstone field derivative terms effectively arising from the Higgs field selfinteraction.
- The neutrino mass is necessarily zero as a consequence of the condensation of the circularly polarized $\vec{\phi}$-modes (for $\bar{h} < h$ and in particular for $\bar{h} = 0$).
- The Higgs sector is larger than in the GWS-model containing an additional Higgs isodoublet field $\varphi'(x)$ carrying fermion number with <u>vanishing</u> ground state expectation value $\langle\varphi'(x)\rangle = 0$.
- There exists an additional isosinglet neutrino which decouples completely from the gauge fields but interacts with the new φ'-Higgs field.
- The Weinberg angle is fixed to $\sin^2\theta_W = \frac{1}{4}$ in lowest approximation.

This looks very encouraging, indeed, but we do have to recall that there are still some distinct differences between the present model Lagrangian and the effective urfield Lagrangian. I have no time to go into further details, except for one remark concerning the Yukawa-type coupling term. Such terms can only be constructed with the urfields if the identification (2.27) for the lepton fields are used. If the quarks are represented as in (2.27) then there exists the possibility of additional Yukawa terms such as to generate also masses for the up-quarks and not only for the down-quarks in contrast to the leptons [18].

5. Quarks and Colour

The most serious problem which remains after this somewhat superficial circumspection is to properly understand colour or whatever it stands for. Formally it requires a further extension of the effective symmetry group. On the basis of our general philosophy we are mainly interested in the question whether colour - similar to hypercharge - can be regarded as a structural off-spring of our basic SU(2,loc) group.

The effective generation of the hypercharge group $U(1,loc)_Y$ from the basic SU(2,loc) was triggered by the condensation phenomenon in the ground state which leads to an iteration of the $U(1,loc)_{I_3}$ subgroup

$$(5.1)\quad \left\{ \begin{array}{l} U(1,loc)_{I_3} \Rightarrow \underset{\downarrow\ (condensate)}{U(1,loc)_{I_3}} \otimes \underset{\downarrow\ (field)}{U(1,loc)_{I_3}} \\ \qquad\qquad\qquad = U(1,loc)_Y \otimes U(1,loc)_{T_3} \\ \text{or } SU(2,loc)_I \Rightarrow U(1,loc)_Y \otimes SU(2,loc)_T \end{array} \right.$$

and establishes an independent conservation law for Y and T for distances $\gg m^{-1}_{Higgs}$.

The question now arises: could there be another mechanism which detaches some dynamical degrees of freedom in such a way as to establish a colour group as an approximately independent group ? It is obvious that the commonly assumed SU(3,loc) colour group cannot possibly occur as an iteration of the basic $SU(2,loc)_I$. The simplest way to achieve this would be to interprete the colour group only as an SO(3,loc) which is homomorphic to SU(2,loc). To accomplish an effective separation one could imagine that there exists the additional possibility of a local or quasilocal condensate, i.e. a soliton-like excitation of the ground state. Such a quasilocal condensate could, in fact, be simply a quasilocal dissolution of the already present, infinitely extended condensate (swiss cheese situation !). The basic symmetry then would effectively decompose into three pieces

$$(5.2)\quad SU(2,loc)_I \Rightarrow \underset{\downarrow}{SU(2,loc)_{\substack{quasilocal\\ condensate}}} \otimes \underset{\downarrow}{SU(2,loc)_{\substack{infinitely\\ ext.\ condensate}}} \otimes \underset{\downarrow}{SU(2,loc)_{field}}$$
$$= SO(3,loc)_{colour} \otimes U(1,loc)_Y \otimes SU(2,loc)_T$$

which would be meaningful for distances $\gg m^{-1}_{Higgs}$. For very small distances flavour-colour transitions of some kind should occur.

One would expect that the quasi-local condensation is not stable by itself but only occurs as an effective dressing of the local fields. (The holes in the swiss cheese form around 'fields'). As a consequence our quasi-local dressing operators constructed from Goldstone fields should be further generalized to include quasi-locally nonvanishing expectation values of the Goldstone field. Such dressings would not lead to a spontaneous breakdown of the corresponding symmetry property (i.e. colour in our interpretation) but rather to a shielding or bleaching of this property. This has still to be investigated in detail. Unfortunately, however, the SO(3) does not appear to suffice completely as a colour group, because in particular

- it does not provide an explanation why only integral charge fields qqq and qq^* (and not also qq^*q^* and qq) should occur,
- it does not permit a hadronic decay of the famous vector resonances ψ, T etc. (G-conjugation type conservation) [25].

The first shortcoming may perhaps be resolved by certain topological properties of the solutions. The second deficiency may not be so bad after all because the hadronic width is experimentally extremely small. The common explanation has here to rely on the Zweig rule and/or a sufficiently small quark-gluon coupling constant. Nevertheless some mechanism is required to explain the hadronic decay of these resonances.

There is some chance that an SU(3) structure may be partially imitated by the fact that the SO(3) symmetry occurs in connection with the nonhermitian isovector-type Higgs fields [20]. With these fields gluon-type gauge fields G_μ and G^A_μ (A=1, ..., 8) can be constructed according to

(5.3)
$$\begin{cases} \vec{\hat{\Phi}} \frac{i}{2} \overleftrightarrow{\partial_\mu} \vec{\hat{\Phi}}^* = \mathbb{1}\, G_\mu(x) + \lambda_A G^A_\mu(x) \\ \vec{\hat{\Phi}} = \vec{\Phi}/\sqrt{\vec{\Phi}^2} \end{cases}$$

which have nonzero 'curvature' $G^A_{\mu\nu} \neq 0$. This situation resembles somewhat the one discussed by Corrigan et al. [26]. The eight gluon fields G^A_μ should actually be interpreted as a triplet and a quintet of SO(3). The additional singlet and quintet lead to a hadronic decay of ψ and T.

It is not clear as yet whether soliton solutions can occur, at all, in the frame work of a spinor theory of our type. Superconductivity teaches us that this will critically depend on the 'Abricosov number', the ratio

(5.4)
$$a = \frac{(\text{penetration length})^2}{(\text{coherence length})^2}$$

which is intimately connected with h/e^2, the ratio of effective Higgs field self-coupling h and gauge field coupling e^2, or to m_{Higgs}/m_{vector}, the ratio of Higgs field mass and mass of the vector gauge fields arising from symmetry breakdown. Superconductivity of the second kind - corresponding to soliton-type solution - occurs if

(5.5)
$$a > a_{crit} \approx .1$$

or in our language if the Higgs mass exceeds the W-mass. This appears reasonable and, in fact, necessary to allow a sufficiently local dressing.

Independent of whether soliton-type solutions are possible or not it is, of course, by no means obvious that they will offer a chance for a dynamical interpretation of the colour property or of whatever property which may equally well explain the observations on hadronic interactions.

As a possible alternative for the interpretation of colour one perhaps should also keep in mind that the urfield dynamics of section 3 does exhibit additional gauge-type interactions connected with spin and spin-isospin rotations for which we have no use up to now.

6. Historical Remarks

In closing my lecture let me make some short historical remarks which relate to the earlier attempts of Werner Heisenberg - in whose memory this Symposium is held - in connection with a unification of elementary particle dynamis. (A more detailed account is given in [20]).

Heisenberg's interest in unification goes all the way back to the late thirties [27] where he points out the possible importance of a new fundamental constant, a universal length, indicating roughly the distance below which the usual quantum mechanical situation should be replaced by something quite different and perhaps reminiscent of a turbulent solution (i.e. soliton-type). He started with theories of the type as suggested by Born [28] in 1933 containing such a universal length and suggested [29] in 1950, in particular, a unified quantum field theory constructed solely from a 4-component self-coupled Dirac spinor field without internal symmetries (nonlinear spinor theory).

Because of the nonrenormalizability of such a theory he studied extensively the quantization of such singular theories and in 1953 pointed out [30] that the quantization of strongly coupled fields has to deviate distinctly from the canonical fields. In this context it became clear that the local field concept is different and more general than the particle concept: Fields parametrize the local dynamics, particles refer to poles in the (asymptotic) S-matrix. This may, for example, imply that local fields are linear operators only in an extended state space with indefinite metric (as in case of Gupta-Bleuler QED) [22,31].

In 1958 Heisenberg and Pauli [32] in an unpublished preprint - after the withdrawal of Pauli finally published in a widely extended form by Dürr, Heisenberg, Mitter, Schlieder and Yamazaki [11] in 1959 - suggested a fundamental field equation for a self-coupled 4-component spinor field which showed a very high symmetry (invariance under 'Touschek transformations' and 'Pauli-Gürsey transformations'). It was shown [33] that this invariance group was an $U(1)_F \otimes SU(2)_I$ and the field actually represented an isospinor-Weyl-spinor field, as discussed here.

To break the symmetry (isospin splitting) and to generate approximate higher symmetries (e.g. the distinction between baryon and lepton number) and anomalous isospin states (hypercharge) an asymmetry of the ground state was postulated [11] (in fact, two years before Goldstone [14] published his important paper). The Heisenberg-Pauli equation (in the interpretation by Dürr) is actually closely related to the urfield Lagrangian introduced here in section 2 if one replaces a factor $(\chi^{*}\bar{\sigma}_{\mu}\chi)(\chi^{*}\bar{\bar{\sigma}}^{\mu}\chi)$ by an effective constant mass square $\mathcal{M}^2$ [21].

In 1961-65 several papers [34] were published on the mass zero problem of the photon, the isospin-anomalous states ('spurion' compounds) and effective higher symmetries. In 1970 Dürr and Winter [17] could resolve the mass zero problem in the conventional fashion after demonstrating that for a dilatation invariant form of the Heisenberg-Pauli equation (appropriate choice of the intrinsic dimension of the urfield) gauge invariance can be established without genuine vector gauge fields (as mentioned in section 2). Saller [35] showed in 1975 that the 'spurions' introduced earlier could actually be constructed in a local fashion from the Goldstone fields and correspond to our quasi-local dressing operators.

There was always much concern about the parity problem in the Heisenberg theory [11,33], i.e. the question whether parity at small distances should become an exact symmetry (and spontaneously broken for large distances) or maximally broken (and then be symmetrized for large distances), a dispute which actually finds its continuation also in the present grand unification models. Starting here with a basic Weyl-type spinor field we actually incorporate locally only PC and <u>not</u> P. This has the advantage of introducing locally only half the number of dynamical degrees of freedom. The observed nearly exact parity symmetry may then appear as a miracle. But for a non-abelian gauge-invariant interaction parity invariance is automatically established if right- and left-handed representations can be locally constructed and belong to <u>equal</u> non-trivial representations of the gauge group.

References

[1] H. Georgi and S.L. Glashow: Phys.Rev. Lett. 32, 438 (1974); A.J. Buras, J.Ellis, M.K. Gaillard and D.V. Nanopoulos: Nucl.Phys. B135, 66 (1978).

[2] H. Georgi: Particles and Fields, 1974 (APS/DPF Williamsburg)(New York, N.Y. 1975), p. 575; H. Fritzsch and P. Minkowski: Ann.Phys. (N.Y.), 93, 193 (1975); M.S. Chanowitz, J. Ellis and M.K. Gaillard: Nucl.Phys. B129, 506 (1977).

[3] F. Gürsey, P. Ramond and P. Sekivie: Phys.Lett. B60, 177 (1976); Y. Achiman and B. Stech: Phys.Lett. B77, 389 (1978); O. Shafi: Phys.Lett. B79, 301 (1978); H. Ruegg and T. Schücker: Nucl.Phys. B161, 388 (1979).

[4] M. Gell-Mann, P. Ramond and R. Slansky: Rev.Mod.Phys., 50, 721 (1978); I. Bars and M. Günaydin: Phys.Rev.Lett. 45, 859 (1980).

[5] S.L. Glashow: Nucl.Phys., 22, 579 (1961); A. Salam and J.C. Ward: Phys.Lett. 13, 168 (1964); S. Weinberg: Phys.Rev.Lett. 19, 1264 (1967); Rev.Mod.Phys. 46, 255 (1974).

[6] P.A. Schilpp: Albert Einstein: Philosopher-Scientist, Tudor Publ.Co., New York, 4th Ed. 1957, p. 20 ff.

[7] W. Heisenberg: Nuovo Cim. 6, 493 (1949); Z.f.Physik, 126, 569 (1949); 133, 65 (1952); Naturwiss. 39, 69 (1952).

[8] H.P. Dürr: Bull.Soc.Math.Belg. 31, 17 (1979).

[9] H.P. Dürr and H. Saller: Phys.Rev. D22, 1176 (1980).

[10] W. Heisenberg: Naturwiss. 63, 1 (1976); H.P. Dürr: Nuovo Acta Leopoldina, 47, 111 (1977).

[11] H.P. Dürr, W. Heisenberg, H. Mitter, S. Schlieder and K. Yamazaki: Z.Naturforschg. 14a, 441 (1959).

[12] H.P. Dürr: Properties of Matter under Unusual Conditions, ed. by H. Mark and S. Fernbach, J. Wiley, New York, 1969, p. 301; Group Theoretical Methods in Physics, Springer Lecture Notes 79, 259 (1977).

[13] H.P. Dürr and H. Saller: Nuovo Cim. A39, 31 (1977); 41, 677 (1977); 48, 505 (1978); 48, 561 (1978).

[14] J. Goldstone: Nuovo Cim. 19, 154 (1961); J. Goldstone, A. Salam and S. Weinberg: Phys.Rev. 127, 965 (1962).

[15] H.P. Dürr and H. Saller: Phys.Lett. B84, 336 (1979); Nuovo Cim. A53, 469 (1979).

[16] H. Umezawa: Nuovo Cim. 38, 1415 (1965); 40, 450 (1965); S. Coleman: Erice Lectures 1973.

[17] H.P. Dürr and N.M. Winter: Nuovo Cim. A70, 467 (1970).

[18] H. Saller: Preprint MPI-PAE/PTh 58/80 Munich, Dec. 1980, to be published in Nuovo Cimento.

[19] H. Saller: Preprint MPI-PAE/PTh 38/81 Munich, July 1981, to be published in Nuovo Cimento.

[20] H.P. Dürr: Heisenbergs einheitliche Feldtheorie der Elementarteilchen, Heisenberg Gedächtnisbuch 1981, Deutsche Akademie der Naturforscher Leopoldina, Dez. 1981.

[21] H.P. Dürr: Nuovo Cim. A62, 69 (1981); Preprint MPI-PAE/PTh 4/81, Munich, March 1981, to be published in Nuovo Cimento.

[22] W. Heisenberg: Nucl.Phys. 4, 532 (1957); W. Karowski: Nuovo Cim. A23, 126 (1974).

[23] H.P. Dürr and H. Saller: Preprint MPI-PAE/PTh 16/81, Munich, March 1981, to be published in Nuovo Cimento.

[24] H.P. Dürr and H. Saller: Nuovo Cim. A60, 79 (1980).

[25] H.P. Dürr, H. Saller and H. Schneider: to be published.

[26] E. Carrigan, D.I. Olive, D.B. Failie and J. Nuyts: Nucl.Phys. B106, 475 (1976).

[27] W. Heisenberg: Annalen der Physik, 32, 20 (1938); Z.f. Physik 11, 241 (1939).

[28] M. Born: Proc.Roy.Soc. (London) A, 143, 410 (1933).

[29] W. Heisenberg: Z.Naturforschg., 5a, 251 (1950).
[30] W. Heisenberg: Nachr. Akad.Wiss., Göttingen, IIa, 111 (1953); Z.Naturforschg. 9a, 292 (1954).
[31] S. Schlieder: Z.Naturforschg. 15a, 448, 460, 555 (1960); K.L. Nagy: State Vector Spaces with Indefinite Metric, Akademiai Kiado, Budapest 1966.
[32] W. Heisenberg and W. Pauli: On the isospin group in the theory of elementary particles, Preprint Jan. 1958, MPI für Physik, Göttingen (unpublished).
[33] H.P. Dürr: Z. Naturforschg. 16a, 321 (1961).
[34] H.P. Dürr and W. Heisenberg: Z.Naturforschg. 16a, 726 (1961); Nuovo Cim. 37, 1446, 1487 (1965); H.P. Dürr and J. Géhéniau: Nuovo Cim. 28, 132 (1963).
[35] H. Saller: Nuovo Cim. A30, 541 (1975); 34, 99 (1976).

SPECULATIONS ABOUT THE QCD VACUUM

Kenneth A. Johnson
Center for Theoretical Physics
Laboratory for Nuclear Science and Department of Physics
Massachusetts Institute of Technology
Cambridge, Massachusetts 02139

1. Introduction

Since quantum chromodynamics is generally accepted as the only serious candidate for the role as the fundamental theory of the strong interactions it is important to try to obtain an understanding of the character of its ground state.

Here I will try to give some indications[1] which suggest that the vacuum state may be described as a quantum liquid of gluon pairs,[2] together with the standard short wave length fluctuations present in the ordinary perturbative vacuum of quantum field theory. The size of the gluon pair wave function will be comparable to the distance between the pairs. There will also be an effective short distance repulsion between the pairs which inhibits their overlap. At the same time since there is no conservation law to prevent the spontaneous creation of the pairs, and we will show that they have a negative energy, all space must be filled with them. It is these features which lead to the liquid character of the state.

In the first part of this talk, I will review the well-known properties of the perturbative ground state in an intuitive way. In the second part, I will discuss the intuitive and semi-quantitative estimates on which a "liquid" description of the ground state is based. In the last part, I will suggest a method on which a more systematic treatment might be based.

2. The Instability of the Perturbative Ground State

In particle language, the standard model is based upon quarks which are color triplets, and gluons which are color octets. Hence, one would expect that the dominant effects should be associated with the gluon-gluon interactions since their charges are the largest. Since QCD is a theory of "charged" massless vector particles one would also expect that magnetic effects are very important, and because the spin of the gluons is one and the quarks one-half, spin also should have the tendency to enhance the importance of the gluon-gluon interaction. We shall

therefore omit the quarks and consider just the gluons, that is, we shall
consider pure QCD.

Recently, it has been pointed out that a simple physical picture of
the origin of the asymptotic freedom of QCD can be given.[3,4] This
picture indicates at the same time the nature of the instability of the
perturbative ground state. For clarity, let us use the example of
electrodynamics. SU(2) gauge theory is the quantum electrodynamics of
massless charged vector mesons. Everything we say may be immediately
extended to any non-abelian theory. The non-abelian local gauge symmetry
follows if the g value of the vector mesons is two, and the meson-meson
coupling constant is equal to the square of the charge. Of course, the
theory is not renormalizable unless this it true. However, the vacuum
polarization calculated to lowest order depends on g being equal to two
but not on the meson-meson coupling constant. Hence to lowest order
asymptotic freedom in the sense of $Z_3 > 1$ must be simply an effect of
spin.

To see this most simply, let us view the perturbative vacuum for any
charged field as a medium of charged particles. We shall for simplicity
always assume that the particles are massless. In the case of fermions,
we have a Dirac sea of negative energy particles. In the case of bosons,
we have a sea of positive energy particles. The vacuum can now be
considered as a polarizable medium which has the special property that
it looks the same in all Lorentz systems. Thus, the product of the
electric permeability, ε, and the magnetic susceptibility, μ, is one;
$\varepsilon\mu = (1/c^2) = 1$. Therefore $\varepsilon > 1$ or "screening" implies $\mu < 1$, or
"diamagnetism." Further $\varepsilon < 1$ or "anti-screening" is equivalent to $\mu > 1$
or "paramagnetism." Since the particles which make up the medium are
charged, when they have a spin they also carry intrinsic magnetic moments.
For renormalizable field theories, in the case of both fermions and
bosons, the g value is two. Since the intrinsic moments are magnetic,
it is simplest to study the polarization properties by considering what
happens in an external magnetic field, rather than an external electric
field. A calculation of the magnetic susceptibility valid for any medium
of free charged particles with spin S_z and g=2 yields, ($\mu=1+\chi$)

$$\chi = e^2 \left\{ Tr\left[(2S_z)^2 - \frac{1}{3} \right] \right\} \times \int \frac{d^3p}{(2\pi)^3} \frac{1}{4E^3}$$

$$\underset{\substack{\text{[Spin}\\ \text{paramagnetism]}}}{} \qquad \underset{\substack{\text{[Landau}\\ \text{diamagnetism]}}}{}$$

where (1)

E=|p| Bosons

=-|p| Fermions

Thus, for spin zero particles the medium is diamagnetic. The Landau
diamagnetism, associated with the quantized orbits of the vacuum particles, corresponds to the standard intuitive picture associated with

charge screening in the case of an applied electric field. In contrast, although it would appear that fermions should be paramagnetic because of the intrinsic magnetic moment carried by the particles the medium carries negative energy. Consequently, spin zero charged particles and spin one-half charged particles both provide a diamagnetic effect and screen but for totally different reasons. The first chance we have for vacuum paramagnetism is with spin one particles; that is, the fact that $\mu>1$ and hence $\varepsilon<1$ for the case of the Yang-Mills theory is a simple consequence of the spin of the particles in association with the positive energy of the vacuum particles.

Since the perturbative vacuum filled with free vector particles is paramagnetic, it can be unstable because the susceptibility increases as the background field strength decreases. The particles can lower their energy below zero. The medium is unstable to spontaneous magnetization. If mean field theory is used to describe the spontaneous polarization of the system of perturbative "free" gluons, we should have taken the first step towards the "Copenhagen" vacuum which is filled with domains of colored magnetic fields.(5) Here, we should like to suggest that the magnetic instability might lead elsewhere.

3. A Vacuum Filled with Gluon Pairs

Quantum QCD is a theory which is characterized by a scale Λ_{QCD} but is without any other parameters. Λ_{QCD} occurs because of the renormalization of the short distance scale fluctuations which can be done perturbatively because of asymptotic freedom. The vacuum could be stabilized classically by the repulsive term A^4 in the Lagrangian, if the energy density locally became large enough (in comparison to $1/\Lambda^4$) to make this the dominant term.

However, we shall argue that there is a possible QCD ground state which is stabilized at the quantum level, that is, with a few field quanta present in the volume $1/\Lambda^3$ where Λ is the basic scale of QCD defined by the short distance perturbative region of the vacuum wave function. We shall not be able to even establish qualitatively that such a quantum vacuum has a lower energy than a possible competing semi-classical state. However, a "shallow" vacuum would seem to be indicated by the success of phenomenological models such as the M.I.T. Bag Model(6) where $B = 55$ MeV/f^3. Here, B is the Bag constant which represents the energy density in the condensed phase (the "outside" vacuum). In terms of the conventional QCD scale Λ, where Λ ranges between 100 to 500 MeV, $B^{1/4}/\Lambda$, ranges between .3 and 1.5. At present the low value is most favored.(7) If we express this in terms of a "number" of gluons with energies $\sim\Lambda$ per volume $(1/\Lambda)^3$, $B\sim n\cdot\Lambda^4$ or $B^{1/4}/\Lambda\simeq n^{1/4}$. Thus, n is not expected to be a very large number. Here we propose to study the "local" stability of the perturbative vacuum, by adding a few localized gluons to it. If this is done throughout the space, we shall then form the "global" state. We shall first look at possible locally stable state with real gluons present in it. We shall then consider the "global" state.

Let us consider first localizing a pair of gluons within a region of space. We shall then show that the pairs exert a strong attraction on each other so that such a pair can be unstable. To describe this in a gauge invariant way we may impose the boundary condition on the field,

$$n_\mu F^{\mu\nu} = 0, \tag{2}$$

where n_μ is the boundary surface. Since

$$\partial_\mu F^{\mu\nu} = g_s i\left[F^{\mu\alpha}, A_\alpha\right], \tag{3}$$

(2) insures that the "color flux", though the surface of the local regior is zero in all gauges. (2) is of course the boundary condition used in Bag Model calculations. Here we consider it more generally as the expression in a gauge covariant way of the localization of quanta. Although we here have a sharp boundary, we shall be interested in applying (2) to the lowest energy states we can achieve within the region. The energy of the lowest states is most sensitive to the overall scale of the region and not to the sharpness of the surface. For simplicity, we shall assume the region to be a sphere. We shall then show that changing the shape will raise the energy. If we quantize the field subject to the boundary condition (2), the lowest modes in a spherical cavity with radius R have energies 2.74/R and total angular momentum and parity 1^+. We may also compute the interaction energy of the gluon pair, which is represented by the diagrams in Fig. 1.

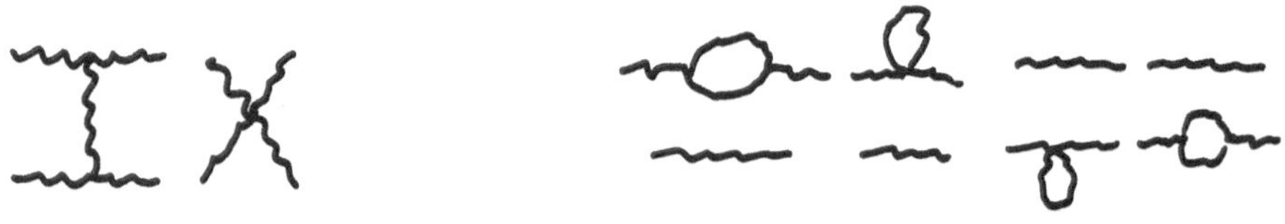

(A) (B)

Fig. 1. Diagrams for <S|H|S> in a state |S> of a gluon pair.

In terms of the color Λ_i and spin S_i matrices of the pair, the energy takes the form

$$-a \frac{\Lambda_1 \cdot \Lambda_2 S_1 \cdot S_2}{R} + b \frac{\Lambda_1 \cdot \Lambda_2}{R} + \frac{1}{2} c \frac{(\Lambda_1 + \Lambda_2)^2}{R} \qquad (4)$$

The terms with coefficients a,b which we shall refer to as the magnetic and electric energies are gauge invariant. The term with coefficient c is gauge dependent. Contributions to the electric energy come from both (A) and (B) diagrams. The magnetic energy in any gauge comes from only the diagrams (A). The coefficient a has been explicitly calculated and has the value $a \simeq .3$. So far, b and c have not been evaluated but an argument has been given which suggests that the numerical value of b may be small in comparison to a, roughly because when both gluons are in the same spatial mode, and form a color singlet, the state is locally color neutral and hence should have a small electric energy. This argument depends very much on the relativistic character of the gluons, that is, that the gluons move so fast that a static color electric field cannot be associated with the relative separation of the pair.*

Let us now consider the magnitude of the magnetic term in the various possible color and spin states for a single gluon pair. The relative values of the strength of this interaction energy are given in Table I. The states are listed in order of the magnitude of the attraction.

Table I

spin	color	$\Omega = -\Lambda_1 \cdot \Lambda_2 S_1 \cdot S_2$
0	1	-6
0	8_S	-3
1	8_A	-3/2
2	27	-1
1	$10, \overline{10}$	0
0	27	+2
2	8_S	+3/2
2	1	+3

*A more familiar way to write the electric term would be in the form $\frac{(b+c)}{R} \Lambda_1 \cdot \Lambda_2 + \frac{1}{2R} c \quad (\Lambda_1^2 + \Lambda_2^2)$. If we were to work in a Coulomb gauge in the case of massive colored particles, b+c would be the electrostatic interaction energy, and the term which comes from the self-interaction diagrams aside from divergent mass renormalization terms would be down by a factor of order $1/m^2R^2$), in comparison. The resulting dominance of b+c would correspond to the presence of a Coulombic color electric field between the particles.

The paramagnetic instability of the perturbative vacuum noted earlier corresponds to the state in which the pair form a vector state which is a color octet, that is the state where $\Omega = -3/2$. If we used mean field theory in this case to describe the global state, we would be at the starting point of the Copenhagen group.

We see a stronger attraction in the spin singlet, color octet channel. In this case a mean field treatment would lead to a Higg's type spontaneous breakdown of color symmetry. However, the attraction is strongest in the spin, singlet, color singlet state, and hence we shall focus here on the possibility that the global state will correspond to a Bose condensed system of spinless and colorless gluon pairs. Condensation in this case would have no symmetry breaking consequences since the condensed particles carry no quantum numbers. Presumably the presence of this matter would strongly modify the effective interaction between colored particles over long distance scales. Here, we shall not discuss whether or not the presence of such a condensate would lead to color confinement.

We shall now show qualitatively how the two gluon states become locally stable, and with a total energy which is negative as a consequence of asymptotic freedom. The local state we have just constructed of course should also contain the "medium" of the perturbative vacuum or at least the high momentum components of it. These gluons modify the coupling between the "real" gluons considered above and cause the coupling parameter α_s to "run" on the scale defined by the basic QCD scale constant Λ, that is, $\alpha_s \to \alpha_s(\Lambda R)$, where for $\Lambda R >> 1$,

$$\alpha_s(\Lambda R) \simeq \frac{1}{A \log(\Lambda R)} \quad , \quad A = \frac{11}{2\pi} \cdot \tag{5}$$

As a consequence the total energy of the gluon pair becomes

$$E(R) \simeq E^{KIN}(R) - (.3)(\alpha_s(\Lambda R))\frac{6}{R} \quad . \tag{6}$$

To be realistic, the estimate of the kinetic energy caused by localizing the pair, $E_{KIN} \sim N2.74/R$, with $N=2$, is considerably too high. If we localize N gluons in a sphere, included in $E_{KIN} = N\cdot 2.74/R$ is not only the cost of the relative localization of the particles which is all that is necessary for them to benefit from the attractive coupling. E_{KIN} also includes a contribution for localizing all the particles together at a fixed place in space (center of "mass" correction). This effect may be simply estimated[8] for a state of N gluons by using the relation

$$E^2_{True} + \langle (p_1 + \ldots p_N)^2 \rangle = (N\cdot\frac{2.74}{R})^2$$

where $\vec{p}_1+\vec{p}_2+\ldots\vec{p}_N = \vec{p}$ is the total momentum of the gluons corresponding to the wave packet described above. Since the gluons are all in the same spacial state,

$$\langle p^2\rangle = N\cdot\langle p_i^2\rangle \simeq N\left(\frac{2.74}{R}\right)^2$$

and therefore

$$E^{KIN}_{True} \simeq \sqrt{N(N-1)}\left(\frac{2.74}{R}\right). \tag{7}$$

Thus, we find for a localized gluon pair,

$$E_{pair} \simeq \sqrt{2}\,\frac{2.74}{R} - 1.8\alpha_s(\Lambda R)\frac{1}{R} \simeq \frac{3.76}{R} - \frac{1.8\alpha_s(\Lambda R)}{R} \tag{8}$$

In (8), the electric energy has been omitted but may be expected to enhance the instability. Because α_s runs, we see that E_{pair} reaches a minimum for $E<0$, and a value of R where $\alpha_s \gtrsim 2.0$. That the local state can be stabilized depends upon asymptotic freedom. The size of the state will be given in units of $1/\Lambda$.

We shall now study the question of whether or not adding more gluons to the localized state lowers the energy still more, that is before we focus on the global form of the ground state, we must study the local state to make sure that it is energetically unprofitable to add more than two gluons locally to the perturbative vacuum. Since gluons are bosons rather than fermions, it is by no means obvious that two gluons will be the most likely configuration locally present in the ground state. We should first consider the limiting case of a state formed with a large number of gluons, N (which for simplicity we shall take to be an even number).

It is not difficult to prove in general for any even number of gluons, all in the same spatial state, and in the spin singlet, color singlet state, that as N becomes large,

$$\Omega_{min.} \rightarrow -3N,$$

where $\Omega = -\frac{1}{2}\sum_{i\neq j}\Lambda_i\cdot\Lambda_j S_i\cdot S_j\ldots_j$. It is also easy to show that the total electric energy is proportional to N. Thus, for large N

$$E^N(R) \simeq N\cdot\left(\frac{2.74}{R} - \alpha_s(\Lambda R)\frac{.9}{R}\right) \tag{9}$$

and hence $E^N(R)$ becomes negative only for a rather larger critical $\alpha_s > 3$. Since this estimate if accurate indicates that N becomes large, and $E<0$, but only in a region where α_s is much larger this calculation suggests that a better procedure would be to look at the local state semi-classically to obtain a self-consistent estimate of the energy. (Again the electric term has been omitted). Perhaps it is this fact that favors the local configuration with a small number of gluons. However, from these considerations alone we cannot judge whether or not the low number of gluon state is energetically favored over a state where a large number of gluons are locally present enough to be described semi-classically. We shall focus on the dilute case because it seems to be indicated phenomenologically.

We should then like to study the stability of the two gluon states against the addition of a small additional number of gluons. The spin and color singlet state of three gluons represents a configuration in which each pair of gluons forms a color octet with spin one, so

$$\Omega = -\Lambda_1\cdot\Lambda_2 S_1\cdot S_2 + \ldots = -3\ (-3/2)(-1) = -9/2. \tag{10}$$

Thus, the total amount of attraction is smaller than in the two gluon states. The first truly competitive situation is that with four gluons. In this case it can be shown that there are four possible spin and color singlet configurations with the values for Ω = -15.97, -4.00, -2.87, +7.85. [The state with maximum attraction may be compared with the large N estimate (-3N=-12) given earlier.] In this case, if we determine the critical coupling at which E<0 we find ($(\alpha_s)_{crit.} \cong$ 2.3) to be compared with ($(\alpha_s)_{crit.} \cong$ 2.0) which was obtained with two gluons. Since these estimates have been made with the electric interaction taken as zero, and the electric interaction should decrease the critical value for α_s in the two gluon case more than in the case for four gluons, we shall tentatively assume that the most stable local configuration is that formed from two gluons.

If the two gluon configurations are the most stable, then even if in the most attractive state, the local energy estimated above the four gluons is close to that of two gluon pairs, we shall argue that there would not be much admixture of the four gluon component locally in the state. This is because if we form a wave function with two pairs when they interact in the S state at small separation most of the time they will find themselves in a state which is much higher in energy than they have when separated. This is because the probability of finding a spin-color singlet pair in the four-gluon state with the maximum attraction, -15.97, has been calculated to be .21. This is the overlap of the two gluon pair state with the most attractive state above. In all the other states, the effective interaction between the two pairs is very repulsive at short distances because the pair of pairs will overlap with one of the other configurations whose value of Ω was listed earlier. Thus, if we form a global state which is densely filled with pairs, in such a state there will be an effective strong repulsion between the pairs which prevents their overlap. In this case we expect that a Bose condensation of gluon pairs of a "liquid Helium type" that is, with a strong inhibition of overlap of the pairs.

These considerations mean that if we were to use a mean scalar field to characterize the pairs, there would be a term $\lambda\phi^4$ in the mean field energy with a rather large coefficient λ. The negative energy in the mean field, $-(M^2/2)\phi^2$ would of course be associated with the negative energy of the isolated gluon pairs.

3. Localized Wave Functions

Another and more basic approach to a microscopic description of the ground state than mean field theory could be based upon the expansion of the gauge field in terms of a complete set of functions $A_k^{n\alpha}(x)$, with two labels n and α, one (n) to characterize the field locally in regions $R\alpha$, and the second (α) used to specify the position in space of the region $R\alpha$. One could then use perturbative methods to approximate the local behavior and thus exploit the benefits of asymptotic freedom. Once the local behavior is controlled, the global label α can be handled with a variational approach. Unfortunately, the simplest way to introduce such a complete set requires the introduction of a regular space lattice which enforces on the system an artificial long range spatial order. Nevertheless, if the ground state has the structure which out earlier considerations suggest that it may, then it might not be completely crazy to begin with such an expansion for the field.

It is not our intention here to go into the details of such an expansion, but only to hint at how it may look.

A one-dimensional example will be sufficient as an illustration of the kind of complete set of functions to be used. Let $u_n(x)$ be any set of functions, complete on the interval (o,a) such that $u_n(o) = u_n(a) = 0$. Define $u_n(x)$ for all x by requiring that $u_n(x+a) = u_n(x)$. The functions $u_n(x)$ may then be non-analytic at the "boundaries" Na, of the intervals into which we have subdivided space, but they are continuous there since $u_n(Na) = 0$. Hence there are no δ-functions in $\partial u_n/\partial x$, and therefore the field energy density will be finite in each mode. Now define

$$U_n^N(x) = u_n(x) \cdot \frac{\sin \frac{\pi}{a}\ (x-(N+\frac{1}{2})a)}{\frac{\pi}{a}(x-(N+\frac{1}{2})a)} \tag{11}$$

The set $U_n^N(x)$ is complete over all x;

$$\sum_{nN} U_n^N(x)U_n^N(y) = \delta(x-y) \tag{12}$$

This can be easily verified using the local completeness of the u's:

$$\sum_n u_n(x)u_n(y) = \sum_N \delta(x-y-Na) \tag{13}$$

One example of a set of u's is $\sin(\frac{n\pi\chi}{a})$. Another set would be $[\cos\frac{n\pi\chi}{a} - 1]$ orthogonalized. The functions U are a special case of functions introduced many years ago in solid state physics and are called "Wannier" functions. In the case of a gauge theory we should like to choose a local set which obey a set of boundary conditions which are guage invariant. An example of such a boundary condition is the one provided by the Bag Model, since it allows one to localize color,

$$n_\mu F^{\mu\nu} = 0. \tag{14}$$

(14) clearly is gauge invariant since $F^{\mu\nu}$ transform homogeneously. For simplicity, let us assume that the normal to the surface is the x direction and let us work in the A = 0 gauge. Then (14) becomes

$$\begin{aligned} Ax &= 0 \\ \frac{\partial Ay}{\partial x} = 0, &\quad \frac{\partial Az}{\partial x} = 0. \end{aligned} \tag{15}$$

Since these conditions are independent of the coupling and they provide a simple way to realize the color localizing type of constraint which we choose to implement in our more intuitive considerations, an expansion of the gauge field in terms of such a complete set of functions would seem to be useful. We are at present trying to develop such a program.

In conclusion, we have indicated how the vacuum might look in terms of a Bose condensed system of bound gluon pairs. Such a "shallow" vacuum might be usefully studied in a microscopic treatment based on the Wannier-functions used in solid state physics.

SOME RECENT PROGRESS IN CHROMO STRING DYNAMICS

J.L. Gervais
Laboratoire de Physique Théorique de l'Ecole Normale Supérieure
Paris - FRANCE

I. INTRODUCTION

So far our belief that Quantum Chromodynamics is the theory of strong interaction is mostly based on the comparison of perturbative calculations with carefully chosen experiments which only test the short distance behaviour of Q.C.D.. The non perturbative predictions of Q.C.D. remain unknown to a large extent although they should reproduce the well-established experimental results on light hadron physics. These notes describe recent attempts to bridge this gap by directly relating Q.C.D. to a dynamics of string similar to that which was found to underlie the dual models. The basic point of this relationship is that the Wilson integrals that gauge invariance forces us to introduce, are functional of curves i.e. have the structure of string field operators as first emphasized in a concrete fashion by Gervais, Neveu(1) Nambu(2) and Polyakov(3).

The work described here, as well as the ideas developed by Makeenko and Migdal(4) tend to show that Q.C.D. is indeed equivalent to a string theory. This string dynamics seems to be different from those derived from the standard models. Hence, it seems appropriate to give it a new name. We shall call it chromo string dynamics or C.S.D. for short.

II. THE LIMIT OF MANY COLOURS

This so-called large N limit, also used in statistical mechanics, was introduced in this problem by 't Hooft(5). Instead of the physical SU_3 colour group, one takes U_N and considers the limit $N \to \infty$. Whether N = 3 can really be regarded as a large number remains an open question, but, for $N \to \infty$ and $g^2 N = cste$ (g is the bare coupling constant), the theory is dominated by the set of all planar Feynman diagrams. These can be characterized by the corresponding two-dimensional surfaces in space-time and, hence, are reminiscent of a string dynamics. This remark was used

References

1. This is a preliminary report of work done in collaboration with T. H. Hansson and C. Peterson to be published more completely later.
2. R. Fukuda, Phys. Rev. D21, 485 (1980) and earlier references contained therein.
3. N. K. Nielsen, Odense preprint, 1980.
4. R. J. Hughes, Phys. Lett. 97B, 246 (1980); R. J. Hughes, Nucl. Phys. B186, 376 (1981).
5. See P. Olesen, Phys. Scripta 23, 1000 (1981) and earlier references contained therein.
6. For reviews, see K. Johnson, Acta Physica Polonica B6, 865 (1975); R. L. Jaffe, 1979 Erice Summer School, MIT CTP-814; P. Hasenfratz, J. Kuti, Phys. Rep. 40C, 75 (1978).
7. See, for example, J. Drees, Report at 1981 International Symposium on Lepton and Photon Interactions at High Energies, Bonn, Aug. 1981.
8. See for example, J. Donoghue and K. Johnson, Phys. Rev. D21 1975 (1980).

This work was supported in part through funds provided by the U.S. Department of Energy (DOE) under contract DE-AC02-76ER03069.

by 't Hooft to postulate that for $N\to\infty$, Q.C.D. becomes equivalent to a string theory.

So far, there is no really convincing proof of this conjecture, but one is getting more and more convinced that it is correct since the developments in recent years strongly support the existence of a chromo string dynamics. The set of planar diagrams in Q.C.D. remains very complicated but great progress has been made in understanding the mechanism of the $N\to\infty$ limit in general[6]. By this, one means that one considers a theory which is invariant under a group, either local or global, with rank N in the limit $N\to\infty$.

Many simpler two-dimensional models can be solved for $N\to\infty$. The simplest is the non-linear σ model (N.L.σ.M.) with Lagrangian

$$\mathcal{L} = -\frac{1}{2}(\partial_\mu \vec{\varphi})^2 \tag{II.1}$$

where space-time is two-dimensional and $\vec{\varphi}$ satisfies the condition

$$\vec{\varphi}^2 = \frac{1}{g^2} \tag{II.2}$$

g is the bare coupling constant and the invariance group is the set of global O(N) rotations of $\vec{\varphi}$. The generating functional

$$Z(J) = \int \mathcal{D}\varphi\, \mathcal{D}\lambda\; e^{\int d_2x \left[\mathcal{L} - \lambda\left(\vec{\varphi}^2 - \frac{1}{g^2}\right) + i\,\vec{\varphi}\cdot\vec{J}\right]}$$

is studied by integrating out $\vec{\varphi}$. This is easy since, after introducing the Lagrange multiplier λ for the condition (II.2), the action has become quadratic in $\vec{\varphi}$. One is left with an effective action for λ which one minimizes to obtain the leading term in the limit $N\to\infty$ with g^2N fixed. This gives, in particular

$$\langle 0|\, \vec{\varphi}(x)\, \vec{\varphi}(y)|0\rangle = \frac{N}{(2\pi)^2}\int d_2k\, \frac{e^{ik\cdot(x-y)}}{k^2+\lambda_c} \tag{II.3}$$

where λ_c is the critical (constant) λ -field.

These explicitly solvable models are nice but for a long time they were solved by special methods, like the one we have just summarized, which does not apply to Q.C.D. More recently, Sakita and Jevicki[7] devised a method which, on the contrary, also works, in principle, for Q.C.D.. The basic idea is that, in general the $N\to\infty$ dynamics can be expressed to leading order entirely in terms of invariant quantities. They introduce a set of such quantities which does not explicitly depend on N and which, although it is over complete for finite N, does provide a complete set of independent invariant quantities for $N\to\infty$. These quantities are functionals of the original fields which they call collective fields. Two examples of such objects are

N.L.σ. M. $$\Phi(x,y) = \vec{\varphi}(x)\cdot\vec{\varphi}(y) \tag{II.4}$$

Q.C.D. $$W[x(\cdot)] = \mathrm{tr}\left[P \exp \oint A_\mu dx_\mu\right] \tag{II.5}$$

These collective fields are introduced as dynamical variables in the description of the system through a change of variables and the $N \to \infty$ limit is obtained by minimizing the corresponding effective action. The solutions, say Φ_0 (N.L.σ.M.) or W_0 (Q.C.D.) thus obtained give the corresponding vacuum expectation values to leading order

$$\Phi_0(x,y) = \langle 0|\vec{\varphi}(x)\,\vec{\varphi}(y)|0\rangle + O(1/N)$$

$$W_0[x(\cdot)] = \langle 0|\,\mathrm{tr}\left[P \exp \oint A_\mu dx_\mu\right]|0\rangle + O(1/N) \tag{II.6}$$

Although it is based on the minimization of an effective action, this method is not a semi-classical approximation to the original problem because one has to take account of the Jacobian of the change of variables which is fully quantum mechanical. For the non-linear σ model, one easily finds back the known solution (II.3), but, unfortunately, the equation for W_0 remains, so far, much too complicated to solve.

Next, another important idea is the concept of master field which was strongly advertized by Witten[8]. The point of $N \to \infty$ limit is that the diagrams with maximal number of closed loops dominate. Hence in the vacuum expectation value of the product of two invariant operators, say G_1, G_2, the dominant diagrams do not contain any propagator connecting G_1 and G_2, so that

$$\frac{\langle 0|G_1 G_2|0\rangle}{\langle 0|0\rangle} \underset{N\to\infty}{\simeq} \frac{\langle 0|G_1|0\rangle}{\langle 0|0\rangle}\,\frac{\langle 0|G_2|0\rangle}{\langle 0|0\rangle} \tag{II.7}$$

Such a general property can be simply understood in the path integral expressions of these vacuum expectation values only if one assumes that the functional integrals are dominated by a single field configuration which is called the master field. This master field is a C number field by construction. It is not however a classical solution of the original field equations because, in the same way as for the collective field recalled above, its dynamics involves an additional quantum mechanical term.

The distinctive property of the master field is that it allows to compute any vacuum expectation of invariant operators from its corresponding C-number expression. For instance in N.L.σ.M. there should exist a field $\vec{\varphi}_M$ such that

$$\frac{\langle 0|G|0\rangle}{\langle 0|0\rangle} \underset{N\to\infty}{\simeq} G(\varphi_M) \tag{II.8}$$

for any O(N) invariant operator G, and this obviously is a solution of equation

(II.7). As a preparation for our later discussion in Q.C.D., let us now determine, following ref. (9), this master field $\vec{\varphi}_M$. It should be such that, in particular

$$\phi_o(x,y) = \sum_a \varphi_M^a(x)\ \varphi_M^a(y) \tag{II.9}$$

where ϕ_o is given by equations (II.6) and (II.3) which can be rewritten as

$$\phi_o(x,y) = \frac{N}{(2\pi)^2}\int \frac{d_2k}{k^2+\lambda_c}\left[\cos(k.x)\cos(k.y) + \sin(k.x)\sin(k.y)\right] \tag{II.10}$$

Equation (II.9) can be looked at as a decomposition of ϕ_o into a sum of a function of x times the same function of y, a requirement which determines φ_M completely. Equation (II.10) is indeed of this form provided one identifies $\sum_a$ with $\int d_2k$ that is if we realize that the $O(\infty)$ group space is a continuous space isomorphic to momentum space. More precisely, equation (II.10) must be regularized and renormalized. If we introduce a space-time box of length L and a momentum cut-off Λ_{max}, we obtain

$$\phi_o(x,y) = \frac{1}{L^2} N \sum_{\substack{n=1,\dots,M\\ m=1,\dots,M}} \frac{1}{k_{nm}^2+\lambda_c}\left[\cos(k_{nm}\cdot x)\cos(k_{nm}\cdot y) + \sin(k_{nm}\cdot x)\sin(k_{nm}\cdot y)\right] \tag{II.11}$$

where k_{nm} has components $\frac{2n\pi}{L}$ and $\frac{2m\pi}{L}$, and

$$\Lambda_{max} = \frac{2\pi M}{L} \tag{II.12}$$

Equation (II.11) is now exactly of the form (II.9) if we set $N = M^2$, $a = (n,m)$. It is obviously simpler and equivalent to introduce a complex master field and to replace (II.9) by

$$\phi_o(x,y) = \sum_{n,m} \varphi_M^{nm}(x)\cdot\varphi_M^{nm}(y)^* \tag{II.13}$$

After a Fourier transform

$$\varphi_M^{nm}(x) = \sum_{\ell,r} \frac{e^{ik_{\ell r}\cdot x}}{L}\ \varphi_{M,\ell r}^{nm} \tag{II.14}$$

we finally obtain

$$\varphi_{M,\ell r}^{nm} = \frac{\sqrt{N}}{\sqrt{k_{nm}^2+\lambda_c}}\ \delta_{n,\ell}\ \delta_{m,r} \tag{II.15}$$

Hence, each component n, m, is a plane wave with the corresponding momentum. When the cut -offs are removed, N goes to infinity and we obtain the continuous $O(\infty)$ group space isomorphic to momentum space.

One sees that in φ_M there is an intimate mixing between space-time and internal space. Lorentz transformations do not leave φ_M invariant since they change momenta. We can however undo this change by applying the inverse transformation to the internal space. These features are reminiscent of monopoles and instantons in Yang-Mills theories. We shall discuss later on about the corresponding structure of Q.C.D. master field..

Lastly, we note that the $O(\infty)$ internal space has to provide a unitary representation of the Lorentz group in view of the above discussion. This group being non compact, such a representation has to be infinite dimensional and can only exist for $N=\infty$.

In general, the master field is not invariant contrary to the collective field. This allows to detect the nature of the $N = \infty$ internal space. On the other hand, this means that the master field is really only defined up to an arbitrary transformation of the internal symmetry group.

III. DUAL MODELS AND STRING DYNAMICS

Before discussing the chromo string dynamics, it is convenient to recall some of the standard properties of the standard dual models(10). In these notes we shall not consider strings with internal degrees of freedom explicitly. Hence, in this paragraph, we restrict ourselves to the Veneziano model. Historically it was obtained as a family of generalized β-functions, satisfying Regge behaviour and duality, all Regge trajectories being parallel to the leading one of equation

$$J = \alpha_0 + \alpha' M^2 \tag{III.1}$$

The model was found to make sense only if $\alpha_0 = 1$ and if the number of space-time dimensions d = 26. Moreover, the underlying physics is the dynamics of a relativistic string. Such a string is described by a field $\phi_\mu(\sigma,\tau)$ which gives the position, at time τ, of the point with curve parameter σ. The time evolution is derived from the Nambu-Goto action which is proportional to the area swept by the string

$$S_{NG} = \frac{1}{2\pi\alpha'} \int d\sigma d\tau \left[(\dot{\phi}_\mu \phi'_\mu)^2 - (\dot{\phi}_\mu)^2 (\phi'_\nu)^2 \right]^{1/2}$$
$$\dot{\phi} \equiv \partial\phi/\partial\tau \qquad \phi' \equiv \partial\phi/\partial\sigma \tag{III.2}$$

Classically, one finds a set of constraints which reflect the invariance of S_{NG} under reparametrization of the surface described by $\phi_\mu(\sigma,\tau)$

$$\mathcal{H}_0 \equiv \frac{1}{2}\left(\mathcal{P}_\mu^2 + \frac{1}{(2\pi\alpha')^2}(\Phi'_\mu)^2\right) = 0 \qquad \mathcal{H}_1 \equiv \Phi'_\mu \mathcal{P}_\mu = 0$$

$$\mathcal{P}_\mu = \delta S_{NG}/\delta\dot{\Phi}_\mu \tag{III.3}$$

The quantization of this action leads to one string states with quantized center of mass position and quantized modes of vibration which reproduce the spectrum of particles predicted by the Veneziano model. In Schrödinger picture these states are described by a wave functional $\Psi[\Phi_\mu(\,)]$ where Φ_μ is an arbitrary function of the curve parameter.

There are two ways to handle the constraints (III.3). First, one can solve them explicitly(11) by using light cone coordinates

$$x^\pm = (x^1 \pm x^0)/\sqrt{2} \qquad \underset{\sim}{x}^i = x^i \quad i = 2,\dots,d.$$

and specifying the parametrization to be such that

$$\Phi^+ = -\tau$$

In this way, one keeps the physical degrees of freedom only (like Q.E.D. in Coulomb gauge) at the expense of using a formalism which is not manifestly Lorentz invariant. A free string in Schrödinger picture is described by a wave functional $\Psi[\tau, p^+, \underset{\sim}{\Phi}(\cdot)]$ where p^+ is the + component of total momentum. It satisfies a Schrödinger equation

$$\frac{1}{2}\int d\sigma \left[-\frac{\delta^2}{\delta\underset{\sim}{\Phi}(\sigma)^2} + \frac{1}{(2\pi\alpha')^2}\underset{\sim}{\Phi}'^2\right]\Psi = i\frac{\partial\Psi}{\partial\tau} \tag{III.4}$$

This method(11) breaks Lorent invariance except if $\alpha_0 = 1$, $d = 26$. Another method of quantization consists in using a wave functional $\Psi[\Phi_\mu(\cdot)]$, that is in keeping all degrees of freedom. Equations (III.3) provide equations which select the physical states. However, in the Fock space of quantized string vibrations these constraint equations have no solution. One, instead, introduces the Fourier components

$$L_m = \int d\sigma \left[\cos(m\sigma) :\mathcal{H}_0: + \sin(m\sigma)\,\mathcal{H}_1\right] \tag{III.5}$$

and (III.3) is replaced by

$$L_m|\Psi\rangle = 0 \qquad m < 0$$
$$(L_0 - \alpha_0)|\Psi\rangle = 0 \tag{III.6}$$

The states satisfying (III.6) are indeed the only ones appearing in the S matrix of Veneziano model. Moreover, they have positive norms is $\alpha_0 = 1$, $d \lesssim 26$. One encounters a similar situation in Q.E.D. where the equation $\partial_\mu A_\mu = 0$ has no solution in the Fock space of photons. One replaces it by $\partial_\mu A^-_\mu|0\rangle = 0$ where A is the annihilation part of A_μ. This last equation does select positive normed states (here for all values of d !).

This last way of quantization is explicitly Lorentz invariant and makes sense for free string at $d \leq 26$. However, it is not more useful than the light cone approach since at $d < 26$ unitarity is broken by the pomeron-like diagrams anyhow. In this connection, Polyakov(12) recently pointed out that the above discussion does not carefully treat the well-known conformal anomaly of conformal invariance which vanishes explicitly at d = 26 only. Whether this really allows to go to $d < 26$ still remains an open question at present.

The field theory of hadrons which is to be related to Q.C.D. is obtained by going to the multi-string formalism where Ψ has become an operator Ψ_{op} which creates and annihilates an entire string. Such an object has been introduced by Kaku and Kikkawa(13) in the light cone formalism where, at the level of free field, Ψ_{op} satisfies (III.4).

IV. THE WILSON LOOP AS A STRING FIELD OPERATOR

As is well-known, gauge invariance forces us to introduce path ordered exponentials and a typical quark-antiquark gauge invariant operator reads

$$\mathcal{M} = \bar{q}(x_1)\left[P\exp \int_{x_0}^{x_1} A_\mu dx_\mu \right] q(x_0) \tag{IV.1}$$

We shall concentrate rather on pure Q.C.D. where the corresponding object is the Wilson loop (II.5) (glueball operator). $W[x(\cdot)]$ is a functional of a path in space-time, a feature which is precisely shared by the functional $\Psi[\phi_\lambda(\cdot)]$ discussed above for dual string quantization. One is thus tempted to consider W as a string field operator. This provides a direct connection between chromodynamics and C.S.D. which is similar to the equivalence between sine-Gordon and massive Thirring models. Precisely, Mandelstam(14) gave an operator proof of this equivalence by building up the Thirring field in terms of the sine-Gordon fields in such a way that the Thirring field equation comes out as a consequence of the sine-Gordon field equations. In Q.C.D. we would thus like to derive the chromo string field equations out of Yang-Mills field equations, a program which was started by Gervais, Neveu(1), Nambu(2).

Typically, a string field equation involves the second functional derivatives with respect to the path. With this motivation, G.N.N. showed that

$$\frac{\delta^2 W}{\delta x_\mu(\sigma)^2} = \mathrm{tr}\, P\left[D_\nu G_{\nu\rho}(\sigma)\, x'_\rho(\sigma)\, \exp \oint A_\lambda dx_\lambda \right] \delta(0) + \mathrm{tr}\, P\left[\left(G_{\rho\nu}(\sigma)\, x'_\rho(\sigma) \right)^2 \exp \oint A_\lambda dx_\lambda \right] \tag{IV.2}$$

$$G_{\mu\nu} \equiv \partial_\mu A_\nu - \partial_\nu A_\mu - [A_\mu, A_\nu]\,; \quad D_\nu G_{\nu\rho} \equiv \partial_\nu G_{\nu\rho} - [A_\nu, G_{\nu\rho}] \tag{IV.3}$$

$$x'_\rho = \frac{dx_\rho}{d\sigma} \tag{IV.4}$$

To relate Q.C.D. to C.S.D. we have to remember that the vacuum expectation value of $W[x(\cdot)]$ is non zero (its behaviour for large loop is the famous Wilson criterion !). Hence, if we define the vacuum expectation values Γ of products of W

$$\Gamma^p[x_1(\cdot), x_2(\cdot), \dots, x_p(\cdot)] = \langle 0| \prod_{\ell=1}^{p} W[x_\ell(\cdot)]|0\rangle \tag{IV.5}$$

we would like to identify the connected parts Γ_c of Γ with the Green functions of C.S.D. and the hope is to use (IV.2) to derive closed equations for Γ_c which would provide the Schwinger-Dyson equations of the chromo string field theory. The kinetic term can always be reduced to $\delta^2/\delta x(\sigma)^2\, \Gamma_c^{p+1}(x(\cdot), \dots)$ with some changes of notations and, following refs. (15), (16), we shall derive an approximate equation for this quantity. First, we want to treat the first term of (IV.2) by using the Heisenberg field equation which will involve the colour currents generated by the Wilson loop. The term to be handled reads

$$\int \mathcal{D}A \exp\left[\frac{1}{4g^2}\int d_4x\,(G_{\mu\nu})^2\right] \mathrm{tr}\, P\left[D_\nu G_{\nu\rho}(\sigma) \exp \oint A_\lambda dx_\lambda\right] \prod_{\ell=1}^{p} W[x_\ell(\cdot)] \tag{IV.6}$$

We do not explicitly write the gauge fixing terms which do not change the conclusion. Formula (IV.6) is transformed by writing

$$e^{\frac{1}{4g^2}\int d_4x\,(G_{\mu\nu})^2}\, D_\nu G_{\nu\rho}(\sigma) = i g\, I^a \frac{\delta}{\delta A^a_\rho(x(\sigma))} \exp\left[\frac{1}{4g^2}\int d_4x (G_{\mu\nu})^2\right]$$

and integrating by parts in the functional integral(17). One obtains

$$-i\int \mathcal{D}A\, e^{-\frac{1}{4g^2}\int d_4x(G_{\mu\nu})^2}\, \mathrm{tr}\left[g\, I^a \frac{\delta}{\delta A^a_\rho(x(\sigma))}\, P\exp \oint_\sigma^\sigma A_\lambda dx_\lambda\right] \prod_{\ell=1}^{p} W[x_\ell(\cdot)] \tag{IV.7}$$

The differential operator $\delta/\delta A_\rho$ acts on all terms on its right and gives a sum of contact terms which can be obtained from the formulae

$$\frac{\delta}{\delta A^a_\rho(x)}\, P\exp\oint A_\lambda dy_\lambda = i g\int d\eta\; y'_\rho(\eta)\, \delta_4(x-y(\eta))\, \mathrm{tr}\left[I^a\, P\exp\int_\eta^\eta d\eta' A_\lambda(\eta') x'_\lambda(\eta')\right] \tag{IV.8}$$

$$\sum_a I^a_{\alpha\beta} I^a_{\gamma\delta} = \delta_{\alpha\delta}\delta_{\beta\gamma} \tag{IV.9}$$

One gets a vanishing contribution except in two cases. If the curve x intersects one of the other curves, say x_ℓ , there is a term where the two Wilson loops are replaced by a single one as in string-string interactions of dual models. Such a term is of higher order in 1/N. If the curve x intersects itself we get a term where W(x) is split into the corresponding product of W over each subloop. To leading order in N^{-1} one of this two W operators is replaced by its vacuum expectation value giving a non local potential for T^{p+1}!

Another point about the first term of (IV.2) is its factor $\delta(0)$ which is ill-defined. We give it a meaning by introducing a cut-off on the number of Fourier modes of $x_\mu(\sigma)$. Since we have closed curves we can write(16),(18)

$$x(\sigma) = \lim_{M\to\infty} x_M(\sigma)$$
$$x_M(\sigma) \equiv \sum_{q=-M',\dots M'} \alpha_q\, e^{2\pi i q\sigma} \tag{IV.10}$$
$$M = 2M'+1$$
$$W[x(\cdot)] = \lim_{M\to\infty} W[x_M(\cdot)] \tag{IV.11}$$

For finite M one can derive(15),(16),(18) an equation similar to (IV.2) where $\delta(0)$ is replaced by M. Moreover, it has been shown(15),(16) that the Wilson loop operator can be renormalized to all orders in perturbation theory for regular curves, hence $W[x_M(\cdot)]$ makes sense and, for finite M, one obtains perfectly well-defined equations.

To leading order in 1/N and neglecting the self intersections, we can drop the first term of (IV.2). The final equation reads

$$\frac{\delta^2 W}{\delta x_\mu(\sigma)^2} = \lim_{M\to\infty} \int d\sigma_1 d\sigma_2\, \delta_M(\sigma_1-\sigma)\,\delta_M(\sigma_2-\sigma).$$
$$\operatorname{tr} P\left[G_{\rho\nu}(\sigma_1)x'_\nu(\sigma_1)\, G_{\rho\lambda}(\sigma_2)x'_\lambda(\sigma_2)\, \exp\oint A_\rho dx_\rho\right] \tag{IV.12}$$

$$\delta_M(\sigma) \equiv \frac{\sin\left[\frac{M}{2}\sigma\right]}{\sin(\sigma)} \xrightarrow[M\to\infty]{} \delta(\sigma) \tag{IV.13}$$

For finite M, the two operators $G_{\rho\nu}x'_\nu$ are now split and therefore this cut-off on the modes has at the same time regularized the operator product $(G_{\rho\nu}x'_\nu)^2$ which appeared in (IV.7).

Next, we discuss how a part of the Q.C.D. spectrum can be approximately derived from equation (IV.12) by using light cone coordinates(15),(16). First, we build the equivalent of dual string operator in light cone formalism by introducing[1]

$$\psi(\tau, p^+, \underset{\sim}{x}(\cdot)) = \int \mathcal{D}x^- \, e^{i\int d\sigma \, p^+ x^-(\sigma)} \, W[x^+ = \tau, x^-(\cdot), \underset{\sim}{x}(\cdot)] \tag{IV.14}$$

As this formula shows, ψ is built out of the Wilson loop operators taken for curves at constant x^+ by Fourier transforming over x^-. One further argues that by reparametrization of the curve one can choose the + component of momentum density to be independent of σ .

In (IV.14) the functional integral of x^- involves highly irregular functions, but for constant x^+

$$\left(x_\mu(\sigma_1) - x_\mu(\sigma_2)\right)^2 = \left(\underset{\sim}{x}(\sigma_1) - \underset{\sim}{x}(\sigma_2)\right)^2 \tag{IV.15}$$

In equation (IV.12) for $M \to \infty$ the two operators Gx' are taken at points with curve parameters $\sigma_1 - \sigma_2 = O(1/M)$. If the curve $\underset{\sim}{x}(\sigma)$ is such that the corresponding points on the curve satisfy

$$\left(\underset{\sim}{x}(\sigma_1) - \underset{\sim}{x}(\sigma_2)\right)^2 \lesssim 1/\Lambda^2_{QCD} \tag{IV.16}$$

where Λ_{QCD} is the typical mass of breaking of scale invariance, it follows from (IV.15) that the product of the two operators Gx' can be estimated from Wilson operator product expansion on the light cone[16]. To leading order equation (IV.12) becomes

$$\left.\frac{\delta^2 W}{\delta x_\mu(\sigma)^2}\right|_{x^+=\tau} = \lim_{M\to\infty} \int d\sigma_1 \, d\sigma_2 \, \delta_M(\sigma_1 - \sigma) \, \delta_M(\sigma_2 - \sigma) \, \langle 0| G_{\rho\nu}(\sigma_1) \, G_{\rho\lambda}(\sigma_2) |0\rangle \, x'_\nu(\sigma_1) x'_\lambda(\sigma_2) \, W \Big|_{x^+=\tau} \tag{IV.17}$$

and we obtain a linear equation which, after Fourier transform over x^- can be cast under the form of a Schrödinger equation similar to (III.4)[15],[16]. A free string dynamics emerges to leading order in N^{-1} as expected.

This chromo string dynamics seems at present to be rather more complicated than the dual string dynamics. Fortunately, we can go further without knowing too much detail about it. For fixed M we have a finite number of variables α_q which we can trade for the curve positions at well-known points[15]

$$\underset{\sim}{x}_s \equiv x_M\left(\sigma = \frac{s}{M}\right) \qquad s = 1, \ldots, M$$

$$\underset{\sim}{x}_s = \sum_{q=-M'}^{M'} \alpha_q \, e^{2\pi i q s/M} \tag{IV.18}$$

This is an orthogonal transformation and, in terms of $\underset{\sim}{x}_s$ we now have a M body problem. For instance, the dual string equation (III.4) leads, after similar discretization to a M body problem with Hamiltonian

$$H = M \left\{ \sum_{s=1}^{M} -\frac{1}{2} \frac{\partial^2}{\partial \underset{\sim}{x}_s^2} + \sum_s V(\underset{\sim}{x}_{s+1} - \underset{\sim}{x}_s) \right\}$$

$$V(x) = \frac{1}{(2\pi\alpha')^2} (x)^2 \qquad \text{(IV.19)}$$

Apart from the factor M, this is a M body Hamiltonian with nearest neighbour interactions. Due to this factor, for $M \to \infty$ (continuum limit), we are only interested in M body low energy excitations (of order 1/M) that is in the phonon spectrum. Q.C.D. leads to a similar Hamiltonian but with a potential of zero range. However, as is often the case in continuum limits (e.g. by fixed points in statistical mechanics) the $M \to \infty$ limit does not depend on the details of V as it only sees the soft modes of the system. This has been argued by Thorn(19) who studied the case of a δ potential. In a R.P.A. approximation due to Goldstone he derived a Regge spectrum with slope (γ = Euler constant)

$$\alpha' = \frac{e^{2\gamma}}{2\pi\sqrt{3}} \frac{1}{\Lambda_s^2} \qquad \text{(IV.20)}$$

where Λ_s is the parameter of scaling violation in the M body problem (there is such a parameter since a δ potential in two dimensions has a dimensionless coefficient).

By the universality argument we are led to apply (IV.20) in chromo string dynamics. Letting $\alpha' \approx 1\ \mathrm{GeV}^{-2}$ gives $\Lambda_s \approx 500$ MeV $\approx \Lambda_{QCD}$ (The discussion of ref. (19) is really for open strings). Λ_s sets the scale of transverse vibrations and if one does not look at high excitations of the transverse modes, (IV.16) is indeed a reasonable approximation.

It is now instructive to see how the information we have just gained on Q.C.D. fits with the results of other approaches. First, one knows, by looking for instance at the Copenhagen vacuum(20), that the Q.C.D. vacuum is very complicated and non-perturbative. It is known in statistical mechanics that in such a disordered situation it pays to introduce a so-called disorder operator. In sine-Gordon, for instance, the description in terms of sine-Gordon fields is good only for small coupling constant K when the fermion is a very massive soliton which does not appear in the vacuum. If K increases, the vacuum becomes populated by soliton-antisoliton pairs (fermion-antifermion) and the good operator to introduce is the Mandelstam operator which creates or annihilates this fermion (disorder operator). In Q.C.D. the coupling constant is not at our disposal but the Copenhagen vacuum is full of flux tubes and the disorder operator is precisely the Wilson loop we have introduced as a dynamical field.

In the vacuum expectation values, W provides colour sources which, due to asymptotic freedom, repel the non perturbative vacuum. Hence, the curve x is surrounded by a tube of perturbative vacuum with width of order Λ_{QCD}^{-1}. If the fluctuations of x are $\lesssim \Lambda_{QCD}^{-1}$ one sees a string structure (fig. 1) and Regge trajectories emerge. This is exactly what we obtained.

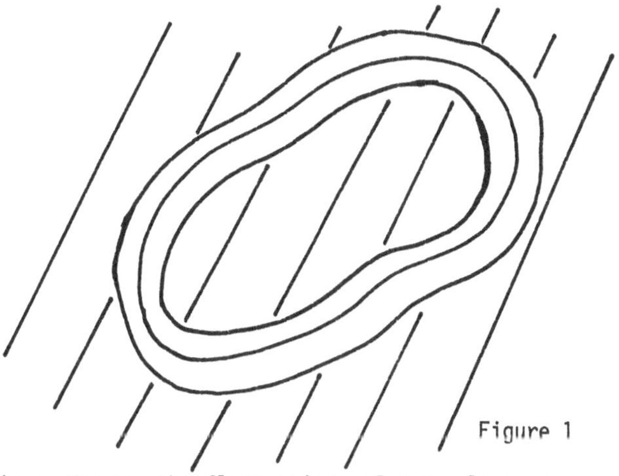

Figure 1

If, on the contrary, the fluctuations of x are large compared to Λ_{QCD}^{-1}, the string structure disappears and the state will look more like a MIT bag (fig. 2).

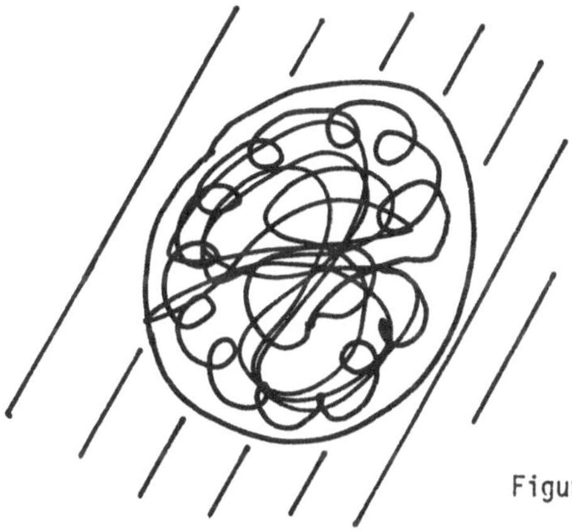

Figure 2

Hence our approach does not conflict with the phenomenological success of this last model.

The last comment is that we have eliminated two degrees of freedom in the chromo string dynamics in exact parallel with dual string dynamics by keeping only the transverse components of the string as a continuous variable. This is an assumption about string dynamics which may be misleading (for instance eliminating two degrees of freedom from a spin 1 particle forces its mass to vanish), but at present our understanding does not go any further.

V. THE MASTER FIELD OF Q.C.D.

In the preceding section we have related the connected Green functions of Wilson loops to the dynamics of the chromo string. On the other hand, these Green functions reduce to their disconnected parts to leading order in N and, following,

Makeenko-Migdal[4] one can study the vacuum expectation value of one Wilson loop to leading order in N. In doing so, one derives properties of the Q.C.D. master field by following the arguments recalled in section II.

The lesson we have learnt from the example of non-linear σ model is two-fold. First, the master field mixes space-time and internal symmetry (In N.L.σ.M., the value of the internal symmetry index was linked to the space-time momentum ; see formula (II.15)). This mixing is reminiscent of solitons, monopoles or instantons. Second, the master field is Lorentz invariant in a peculiar way : a Lorentz transformation, which scales momenta, must be accompanied by an internal symmetry transformation, to restore the initial values of the internal space indices. This argument is of course highly formal, since the Lorentz group is non-compact, contrary to the internal symmetry group, except at $N\to\infty$. The original theory requires, at the beginning, ultraviolet and infrared cut-offs in which case one has a master field with a finite number of components, that is finite N. Due to the mixing between space-time and internal space the removal of cut-offs and the limit $N\to\infty$ are linked in a precise way and one can hope to achieve Lorentz invariance only at $N\to\infty$. This formal realization of the Lorentz group in the internal symmetry space is probably quite general, and reminiscent of the realization of the rotation group for the 't Hooft-Polyakov monopole.

An example of a solution to the classical four-dimensional Yang-Mills equations that exists only at $N=\infty$, has been found by Banks and Casher[21] and called the indexon. In this clever example, the internal symmetry space of $N\to\infty$ is identified with ordinary space-time. However, the indexon does not give an area law for the Wilson loop. In our opinion,[22] this failure is due to the fact that the internal space of the indexon is still too small.

Indeed, as we already discussed, the invariant quantities to be considered for $N\to\infty$ in Q.C.D. are the Wilson loops which are functionals of curves. By analogy with non-linear σ model we can expect the internal space of $N\to\infty$ to be of the same nature as the Wilson loop space, namely to be a space of functionals of curves.

This can be made more precise[22] by looking at the equation derived by Migdal and Makeenko[4] for the $N\to\infty$ limit

$$\frac{1}{N}\langle 0|\operatorname{tr} P\left(D_\nu G_{\nu\mu}(x)\exp\int_{C_{xx}} A_\rho dx_\rho\right)|0\rangle$$

$$= \int_C dy_\mu\,\delta(x-y)\,\langle 0|\frac{1}{N}\operatorname{tr} P\exp\int_{C_{xy}} A_\rho dx_\rho|0\rangle\langle 0|\frac{1}{N}\operatorname{tr} P\exp\int_{C_{yx}} A_\rho dx_\rho|0\rangle \tag{V.1}$$

where C is an arbitrary curve.

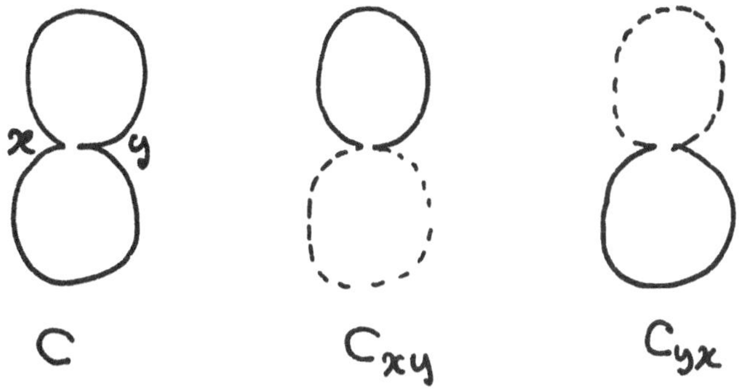

Figure 3

To derive this relation one uses equations (IV.8) (IV.9) in the case of a single Wilson loop operator and keeps only the leading term in N^{-1}. This equation is thus valid to leading order in 1/N, where both sides, being gauge invariant, are simply given by the corresponding value of the functionals, evaluated at the master field $\tilde{A}$

$$\frac{1}{N} \mathrm{tr}\, P\left[D_\nu \tilde{G}_{\nu\mu}(x) \exp \int_C \tilde{A}_\rho dx_\rho \right]$$

$$= \int dy_\mu\, \delta(x-y) \frac{1}{N} \mathrm{tr}\left[P \exp \int_{C_{xy}} \tilde{A}_\rho dx_\rho \right] \frac{1}{N} \mathrm{tr}\left[P \exp \int_{C_{yx}} \tilde{A}_\rho dx_\rho \right] \tag{V.2}$$

These equations can be considered as a set of linear equations for $D_\nu \tilde{G}_{\nu\mu}$ at some fixed x, one for each curve C going through x. If there were more equations than matrix elements of $D_\nu \tilde{G}_{\nu\mu}$, (V.2) would imply that there are constraints between Wilson loops for different curves, at $N = \infty$. This contradicts current belief if $\tilde{A}$ is sufficiently generic. Hence we conclude that $D_\nu \tilde{G}_{\nu\mu}$ itself must be an operator acting in a loop space. We are thus led to the idea that the master field of $U(\infty)$ Yang-Mills should have internal symmetry indices lying in a functional space, the space of configurations of a string with suitable boundary conditions. Lorentz transformations will of course be accompanied by suitable compensating transformations on the internal string configuration space.

Next we discuss how one can define a gauge field where the internal space of $U(\infty)$ Yang-Mills is the space of all configurations of a string. This means that this vector potential which is a matrix in internal symmetry space, becomes an operator defined on this space, and multiplying such operators automatically involves a functional integration over the string configurations in the intermediate states. Taking a trace is also performing a functional integration.

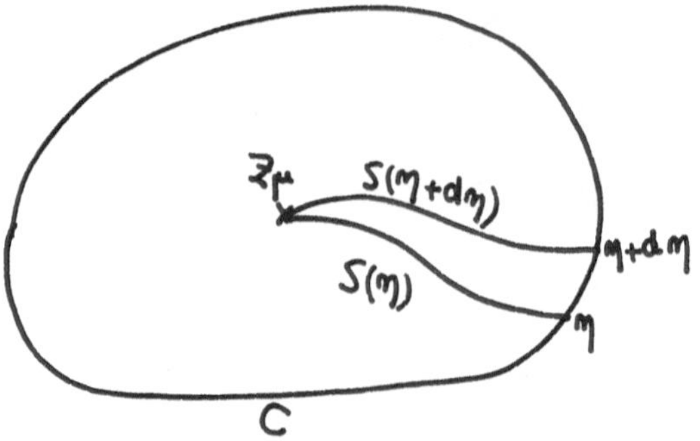

Figure 4

To build a confining ansatz for U(∞) Yang-Mills, we consider fig. 4. C is a loop, which lies in ordinary space-time, and is defined by four functions $X_\mu(\eta)$ (up to reparametrizations). z_μ is also a point in space-time, and $S(\eta)$ and $S(\eta+d\eta)$ two string configurations running from z to the points X and $X+dX$ on the loop. Although the four coordinates of X and the string configurations $S(\eta)$ and $S(\eta+d\eta)$ are in space-time, we shall use them, as announced in the introduction, as internal space labels. As the point X runs along the Wilson loop, it obviously generates, by dragging along the attached string $S(\eta)$, a surface, parametrized by the functions $\phi(\sigma,\eta)$ where σ parametrizes a given string $S(\eta)$. As an ansatz for the Wilson loop, we shall take the functional integration over all such surfaces, at fixed C :

$$\int \mathcal{D}\phi(\cdot,\cdot)\, \exp\left[\frac{-1}{2\pi\alpha'}\int d\sigma d\eta \left[\left(\frac{\partial\phi_\mu}{\partial\sigma}\frac{\partial\phi_\mu}{\partial\eta}\right)^2-\left(\frac{\partial\phi_\mu}{\partial\sigma}\right)^2\left(\frac{\partial\phi_\nu}{\partial\eta}\right)^2\right]^{1/2} \tag{V.3}$$

In the exponent, we have used the Nambu-Goto string action. α' is an arbitrary number, with dimensions $(\text{mass})^{-2}$, which we cannot determine for the moment. In the remaining part of the lecture, we shall discuss this ansatz, the internal space structure that it implies, and the induced behaviour of the Wilson loop.

At this level of the discussion, we shall not worry about problems arising from the reparametrization invariance in σ and η of the ansatz (V.3). If we have a reasonable ansatz for the Wilson loop, it must be possible to rewrite it in the traditional form

$$\mathrm{tr}\left[P\exp\int \tilde{A}_\mu dx_\mu\right] \tag{V.4}$$

i.e. it should be possible, via matrix multiplication in internal space, to glue two pieces of the loop together, and the operator $(1+\tilde{A}_\mu dX_\mu)$ will just be the effect of adding the infinitesimal piece dX_μ . $\tilde{A}_\mu$ thus defined must not depend on dX , only on X , and on the initial and final string configurations $S(\eta)$ and $S(\eta+d\eta)$. The matrix multiplication becomes a functional integral over the function ϕ_μ of σ (at fixed η) which describes the string $S(\eta)$. This factorization property has precisely been proved in ref. (23) for the linearized form of the Nambu-Goto action, which is obtained in dual-resonance models by choosing orthogonal coordinates on the surface $\phi_\mu(\sigma,\eta)$ and in the particular case of Z sitting on the curve C. Formally, for any Z, it is a simple consequence of the additivity of the classical action and of the composition law of Feynman path integrals.

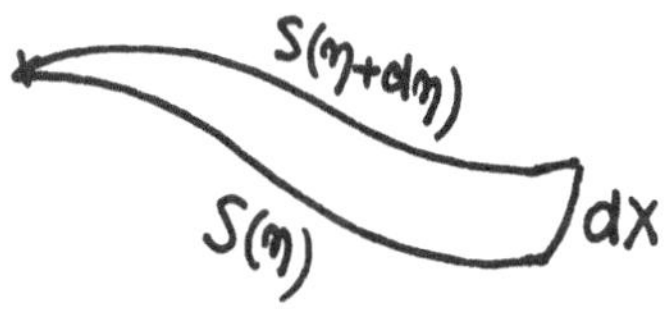

Figure 5

$(1+\tilde{A}_\mu dX_\mu)$ is associated with the infinitesimal wiggly triangle of fig. 5 and formula (V.3) corresponds to the sewing together of these triangles, to build up the entire surface. We shall come back to the actual derivation of $\tilde{A}_\mu$ in a future publication. The main problem in a correct treatment of reparametrization invariance. We deal with it in ref. (22). Albeit for different boundary conditions on we obtain another example of an $U(\infty)$ gauge field with functional internal space.

In the ansatz described by fig. 4, the internal space indices are thus the four coordinates z_μ and the degrees of freedom of the string S, which as boundary conditions, must go from z_μ to X_μ when σ goes from 0 to π (say). Such a set of degrees of freedom can be described by a discrete infinite set of Fourier modes $a_{m\mu}$ obtained by expanding the function ϕ_μ which describes the curve S

$$\phi_\mu(\sigma,\tau) = \left(1-\frac{\sigma}{\pi}\right) z_\mu + \frac{\sigma}{\pi} X_\mu + \sum_{m=1}^{\infty} a_{m\mu} \sin(m\sigma) \tag{V.5}$$

Because the exponent in (V.3) contains the area of the surface spanned by the curve $S(\eta)$, we can expect that large Wilson loops will behave as exp $-(\frac{1}{2\pi\alpha'}$ times the area of the minimal surface), leading to confinement.

Just as in the non-linear sigma-model, a mass scale, α' , appears in the master field. Its precise value will come out of dimensional transmutation when the $N\to\infty$ limit is finally understood.

Going back to the area law, one sees that the details of what happens for smaller loops will depend on the way one breaks the reparametrization invariance of (V.3). Breaking this invariance is necessary to define the functional integral of (V.3). The conventional dual-resonance treatment gives a well-defined answer to this problem, which, as we recalled in sect. 3, is Lorentz invariant only for a certain value of the dimension of what is now the internal space. This so-called critical dimension, rather than being an embarassement can be used to specify the internal space of $N \to \infty$ limit. The fact that it is larger than four causes no problem since, in formula (V.3), space-time is only at the boundary.

In section IV, we connected chromo string dynamics with the connected Green functions Γ_c^p which, to leading order in N^{-1} describe the set of connected planar diagrams with $p \geq 2$ holes. The master field describes the set of planar diagrams with one hole. It is likely that the correspondence between planar diagrams and chromo string dynamics is local on the corresponding fluctuating two-dimensional surfaces and the master field is thus an operator in chromo string Hilbert space. This viewpoint sheds new lights on the recent works by Migdal (24) which we now briefly recall. This author introduces a quantum theory of surfaces ϕ_μ, with a prescribed boundary similar to the one we used together with two dimensional Fermi fields $\psi \bar{\psi}$. Their ansatz for the Wilson loop is, instead of (V.3)

$$Z[C] = \int \mathcal{D}\phi \, \mathcal{D}\psi \, \mathcal{D}\bar{\psi} \; e^{-\int d\sigma d\eta \left[\bar{\psi} \gamma \cdot \partial \psi + m g^{1/4} \bar{\psi} \psi \right]} \tag{V.6}$$

where g is the determinant of metric on the surface. Migdal (24) has gone quite some way towards showing that such an ansatz satisfies equation (V.1) where the left member is replaced by a second order functional derivative with respect to the curve. For us[22], formula (V.6) defines another ansatz for the masterfield. This masterfield involves fermionic degrees of freedom in addition to the orbital string modes. One may consider the work of ref. (24) as an attempt to show that (V.6) indeed provides the master field of Q.C.D. It is however based on the use of a functional derivative which seems delicate to handle in quantum field theory. Our work suggests a simple procedure which could be to compute the master field $\tilde{A}$ from our ansatz and try to verify (V.1) directly by computing $D_\nu \tilde{G}_{\nu\mu}$.

Once the master field is found, one can study the corresponding string dynamics in space-time, which, as was proposed by G.N.N.[1],[2], is done by applying standard functional derivatives to Green functions of Wilson loops. The simplest such Green function is of course given by our ansatz for large N and this space-time dynamics will naturally emerge to be of the string type in view of the string structure of internal space and the way it mixes with space-time. Hence, the master field provides yet another strong indication for the existence of chromo string dynamics.

REFERENCES

(1) J.L. Gervais, A. Neveu, Phys. Lett. 80B (1979) 255
(2) Y. Nambu, Phys. Lett. 80B (1979) 372
(3) A.M. Polyakov, Phys. Lett. 82B (1979) 247
(4) Y.M. Makeenko, A.A. Migdal, Phys. Lett. 88B (1979) 135
(5) G 't Hooft, Nucl. Phys. B75 (1974) 461
(6) For a recent pedagogical review, see S. Coleman : "1/N" lecture at the 1979 Erice Summer School
(7) B. Sakita, Phys. Rev. D21 (1980) 1067
A. Jevicki and B. Sakita, Nucl. Phys. B165 (1980) 511, Phys. Rev. D22 (1980) 467
(8) E. Witten, Harvard preprint and Cargèse lecture in Recent developments in gauge theories" 1980 Plenum Press,see also ref. (6)
(9) A. Jevicki and H. Levine, Phys. Rev. Lett. 44 (1980) 1443
(10) for reviews see : "Dual theory" edited by M. Jacob, North Holland ; J. Scherk Rev. Mod. Physics 47 (1975) 123
(11) P. Goddard, J. Goldstone, C. Rebbi, C.B. Thorn, Nucl. Phys. B56 (1973) 109
(12) A.M. Polyakov, Phys. Lett. 103 B (1981) 207
(13) M. Kaku, K. Kikkawa, Phys. Rev. D 10 (1974) 1110 and 1823
(14) S. Mandelstam, Phys. Rep. 23C (1976) 307
(15) J.L. Gervais, A. Neveu, Nucl. Phys. B163 (1980) 189
(16) Pedagogical lectures describing my works with A. Neveu are :
"String states in Q.C.D." Cargèse lecture, "Recent developments in gauge theories" Plenum Press 1980
"Q.C.D. strings" Dubrovnic lecture in "Particle Physics 1980" North Holland
"Strings in Q.C.D. with many colours" lecture at XIIth seminar on theoretical physics (G.I.F.T.) 1981
(17) A.A. Migdal, An. of Phys. 126 (1980) 279
(18) J.L. Gervais, A. Neveu, Nucl. Phys. B153 (1979) 445
(19) C.B. Thorn, Phys. Rev. D19 (1979) 639
(20) for a pedagogical review see : H.B. Nielsen, Dubrovnik lecture "Particle Physics 1980" North Holland
(21) T. Banks, A. Casher, Nucl. Phys. B167 (1980) 215
(22) J.L. Gervais, A. Neveu, LPTENS preprint 81/6, to be published in Nucl. Phys. B
(23) J.L. Gervais, B. Sakita, Phys. Rev. D4 (1971) 2291
(24) A.A. Migdal, Phys. Lett. B96 (1980) 233, Landau Institute Preprint "Q.C.D.-Fermi string theory"

THE NUMERICAL STUDY OF QUANTUM CHROMODYNAMICS†

Julius Kuti*

Institute for Theoretical Physics
University of California
Santa Barbara, California 93106
and
Center for Theoretical Physics
Laboratory for Nuclear Science and Department of Physics
Massachusetts Institute of Technology
Cambridge, Massahcusetts 02139

ABSTRACT

A powerful stochastic method is presented for the numerical evaluation of path integrals in quantum mechanics. The method is directly applicable for the detailed numerical study of Quantum Chromodynamics (QCD) with lattice regularization. Important results on non-perturbative physical quantities, like the confining force between a heavy quark-antiquark pair, the critical temperature of thermal quark liberation, or the mass gap for glue-ball excitations are reviewed first, within the pure gauge sector of the theory. The stochastic treatment of the complete fermionic problem is also described.

---

†This presentation is based on invited talks delivered at the Heisenberg Symposium in Munich, July 1981, and the 1982 Annual Joint APS/AAPT Meeting in San Francisco, January 1982.

*On leave from the Central Research Institute for Physics, Budapest.

The outline of this presentation is as follows:

1. Introduction

2. The stochastic method
   (a) Feynman's path integral
   (b) the Markov process
   (c) particle in double well potential

3. The pure gauge sector of QCD
   (a) string tension
   (b) critical temperature of thermal quark liberation
   (c) mass gap and glue-ball excitations
   (d) gluon condensate in the vacuum

4. Stochastic treatment of the fermion problem
   (a) fermion determinant
   (b) hadron spectrum without internal quark loops
   (c) method to treat the complete fermionic theory

1. Introduction

After so many years there is now a general belief that we have found the theory of strong interactions as described by Quantum Chromodynamics.[1] This theory is remarkably consistent with deep inelastic phenomena and provides now a consistent framework for probing hadron structures at short distances.

The great success of QCD is explained by asymptotic freedom[2] which tells you that the quark-gluon coupling becomes weak at short distances and perturbation theory is applicable.

On the scale of 1 fermi the coupling becomes strong, perturbation theory breaks down and the most interesting hadronic properties (mass spectrum, decay widths, wave functions, . . .) are protected from analytic calculations.

In fact, we know on quite general grounds that any hadron mass M_H depends on the bare coupling constant g(a) as

$$M_H = \text{const}\cdot a^{-1} e^{-\frac{c}{g^2(a)}} \tag{1.1}$$

which renders perturbation theory useless on the fermi scale. The space-time cut-off a at short distances carries the dimension of the mass in Eq. (1.1). It is the only dimensional quantity in the regularized theory in the limit when all quark masses vanish. The constant C in the exponent is calculable analytically, but the overall constant in front of the singular exponential function in Eq. (1.1) is not accessible by analytic methods.

The bare coupling constant g(a) is a logarithmic function of the cut-off a,

$$g(a) \sim \frac{1}{\ln \Lambda a} \quad , \tag{1.2}$$

as known from asymptotic freedom. The scale parameter Λ determines the strength of the quark-gluon coupling in QCD. Now, we note that the expression $a^{-1}\exp(-c/g^2(a))$ in Eq. (1.1) does not vary as we change the cut-off at very short distances.

Through this dimensional transmutation of Eq. (1.1) any hadron mass is of the form

$$M_H = \text{const}\cdot \Lambda_{\overline{MS}} \tag{1.3}$$

where $\Lambda_{\overline{MS}}$ is the experimentalist's favorite scale parameter.[3] The constant in front of $\Lambda_{\overline{MS}}$ in Eq. (1.3) must come from a non-perturbative calculation.

The numerical calculations I will present here strongly indicate a value around $\Lambda_{\overline{MS}} \sim 100$ MeV in the pure gluon sector of the theory, with perhaps about twenty percent uncertainty. If this value of $\Lambda_{\overline{MS}}$ will not change significantly in the presence of virtual quark vacuum polarization, SU(5) may run into trouble with the lifetime prediction[4] for proton decay,

$$\tau_p \sim 10^{32\pm 2} \left(\frac{\Lambda_{\overline{MS}}}{GeV}\right)^4 \; yr. \qquad (1.4)$$

If one wants to interpret the recent underground Tata experiment in India as a lower bound,

$$\tau_p > 8 \cdot 10^{30} \; yr \, ,$$

then the situation is already becoming tight with $\Lambda_{\overline{MS}} \sim 100$ MeV.

The value of $\Lambda_{\overline{MS}}$ is only poorly known from short distance physics to be somewhere between 100 MeV < $\Lambda_{\overline{MS}}$ < 400 MeV. The main difficulty is that the predicted scaling violation effects are logarithmically slow functions of the momentum transfer. At present, experimentalists tend to prefer $\Lambda_{\overline{MS}}$ values in the lower range around 100-200 MeV.

In our numerical study of QCD we hope to determine $\Lambda_{\overline{MS}}$ accurately and less costly. We have a better chance for good accuracy, since the important relations (like Eq. (1.3)) are linear in $\Lambda_{\overline{MS}}$. It would be a great triumph to calculate the constant in Eq. (1.3) within twenty percent accuracy for the proton, the pion and other non-perturbative quantities.

The method I am going to describe is in a way an experiment. But, very importantly, it is our experiment. We, theorists, can create now interesting conditions inside the computer like the liberation of quarks at high temperature, or a container filled with glueball gas.

To understand the method, I describe it first in detail as applied to a simple quantum mechanical problem. The exciting results in Quantum Chromodynamics I will only summarize without going into details. This presentation is prepared for those who want to get a first idea about the method and who also want to see the list of results accomplished so far in Quantum Chromodynamics.

2. The stochastic method

I will briefly describe a powerful stochastic method for the numerical evaluation of path integrals in quantum mechanics.[5] The method is easy to generalize to Quantum Chromodynamics with lattice regularization.

Consider a non-relativistic particle of mass m moving in one dimension in a potential well V(Q). The particle is described by the Hamiltonian

$$H = \frac{P^2}{2m} + V(Q) \quad . \tag{2.1}$$

The transformation matrix element (transition amplitude)

$$Z_{fi} = \langle Q_f | e^{-\frac{i}{\hbar} H(t_2 - t_1)} | Q_i \rangle \tag{2.2}$$

plays a fundamental role in quantum mechanics. It describes the probability amplitude of propagation from the initial position Q_i of the particle at time t_1 to some position Q_f at time t_2.

2a. Feynman's path integral

The time interval $t_1 - t_2$ in Eq. (2.2) can be divided into n + 1 segments with $t_2 - t_1 = (n + 1)\varepsilon$, where ε is the time slice on the time lattice of Fig. 1.

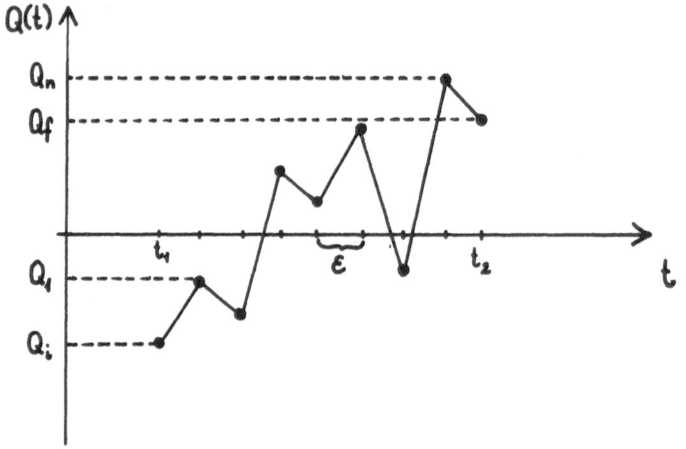

Fig. 1

It is easy to show that Z_{fi} in Eq. (2.2) can be represented as the limit of the n-dimensional integral,

$$Z_{fi} = \lim_{n \to \infty} \int \prod_{k=1}^{n} \frac{dQ_k}{\left(\frac{2\pi\hbar i\epsilon}{m}\right)^{\frac{1}{2}}} \exp\left\{\frac{i}{\hbar} \sum_{j=1}^{n} \epsilon \left[\frac{1}{2} m \frac{(Q_j - Q_{j-1})^2}{\epsilon^2} - \frac{V(Q_j)+V(Q_{j+1})}{2}\right]\right\}$$

with the notation $Q_0 = Q_i$ and $Q_{n+1} = Q_f$. (2.3)

The integration over n variables in Eq. (2.3) corresponds to the sum over all zig-zag paths of Fig. 1 connecting Q_i to Q_f in the time interval $t_2 - t_1$. The limit $n \to \infty$ defines Feynman's path integral,

$$Z_{fi} = \int \left[\frac{dQ}{\left(\frac{2\pi\hbar i\epsilon}{m}\right)^{\frac{1}{2}}}\right] \exp\left\{\frac{i}{\hbar} \int L(Q,\dot{Q})\, dt\right\} , \qquad (2.4)$$

where $L = (1/2)m\dot{Q}^2 - V(Q)$ is the Lagrangian of the particle. The transition amplitude Z_{fi} $(Q_f = Q(t_2), t_2; Q_i = Q(t_1), t_1)$ is given by the sum of phases along all paths from Q_i to Q_f. The phase is determined by the action $S = \int L(Q_1, \dot{Q})dt$ along the path.

For the numerical evaluation of the path integral we will keep n finite and work on a time lattice as depicted in Fig. 1. With sufficiently dense slicing (large n) the calculation of the path integral becomes accurate.

It is very difficult to sum the rapidly oscillating phases of the path integral, even in simple quantum mechanical applications. We follow the standard trick and rotate to Euclidean time τ with $t = i\tau$. The rotation makes real Minkowski time t now purely imaginary. In practice, this rotation corresponds to the replacement $\epsilon = ia$ in the zig-zag approximation of Eq. (2.3). The Euclidean time slice a is a small positive number on our Euclidean time lattice.

The Euclidean transformation matrix element with n moving points on the zig-zag paths is given by

$$Z_{fi}^{(n)} = \int \prod_{k=1}^{n} dQ_k \exp\left(-\frac{1}{\hbar} S_E(Q_i, Q_1, \ldots, Q_n, Q_f)\right) \qquad (2.5)$$

where we dropped the normalization factor of the integrand. It always cancels in physical quantities.

In this approximation the Euclidean action S_E is given by

$$S_E(Q_0, \ldots Q_{n+1}) = \sum_{k=1}^{n} \left[\frac{1}{2} ma \frac{(Q_{k+1} - Q_k)^2}{a^2} + a\frac{V(Q_{k+1}) + V(Q_k)}{2}\right] \qquad (2.6)$$

and there is no integration in Eq. (2.5) over the end points $Q_0 = Q_i$ and $Q_{n+1} = Q_f$ of the Euclidean time interval.

One can study the correlation functions of the system in Euclidean time, like the two-point function

$$\frac{1}{Z_{fi}^{(n)}} \cdot \int \prod_{k=1}^{n} dQ_k \cdot Q_l \cdot Q_{l'} \exp\left(-\frac{1}{\hbar} S_E(Q_0, \ldots, Q_{n+1})\right), \tag{2.7}$$

or more complicates ones. The relevant physics has to be extracted from the correlation functions.

2b. The Markov process

Feynman's path integral formulation in Euclidean time has a close analogy with a classical one-dimensional crystal. In the approximation of n integration variables, $Z_{fi}^{(n)}$ in Eq. (2.5) can be regarded as the partition function of a chain of n particles with fixed end points. There is some on-site potential energy $aV(Q_k)$ for each displacement Q_k, and there is some nearest neighbor interaction $(m/2a)(Q_{k+1} - Q_k)^2$. The energy of this classical system is given by the Euclidean action S_E. The "temperature" of the crystal is $\hbar$ formally.

Also, the Euclidean Green's functions of the original quantum mechanical problem are in one to one correspondence with the correlation functions of the one-dimensional crystal.

We want to calculate now multi-dimensional integrals of the type as in Eq. (2.7) by importance sampling of a stochastic procedure. Since the algorithm carries the awkward and misleading name of Monte Carlo method, experimentalists might get the wrong impression that they can also run this experiment.

The secret of importance sampling is that instead of throwing points into the large phase space of the n integration variables, we try to generate a distribution according to the most rapidly varying part of the integrand which is $\exp\{(-1/\hbar)S_E\}$ in our case. This is implemented through a Markov chain.

A point $(Q_1 \ldots Q_n)$ in the n-dimensional configuration space represents a state of the crystal. We shall define now a stochastic process which generates new states from the initial one, step by step along a Markov chain in the state space of the system.

We start from an arbitrarily chosen point Q in the configuration space and cycle through the crystal from site to site making only a local change in the state of the crystal when we stay on a given site. On a site with some label k we generate a new value of the variable Q_k according to the probability distribution $\exp\{(-1/\hbar)S_E\}$ where all other variables are kept fixed. S_E is regarded on site k as a function of Q_k only.

This way of cycling through the lattice corresponds physically and intuitively to touching a heat bath of the right temperature $\hbar$ to the sites of the crystal, step by step in a sequential manner. Each new state after the up-date of a site is the next element of the Markov chain.

After many sweeps we bring the crystal to thermal equilibrium at temperature $\hbar$ and the elements of the Markov chain generate now a distribution of states in configuration space according to the desired distribution $\exp\{(-1/\hbar)S_E(Q_1 \ldots Q_n)\}$ in all variables.

2c. Particle in double well potential

This simple application of the above described method will demonstrate the power of the stochastic procedure. It will also help us to understand the thermodynamics of a quantum mechanical particle in a heat bath at termperature T using our formulation of the problem.

In thermodynamics the partition function of the particle is given by

$$Z = \mathrm{Tr}\, e^{-\beta H} \quad , \quad \beta = \frac{1}{kT} \tag{2.8}$$

which can be rewritten as

$$Z = \int dQ \langle Q | e^{-\beta H} | Q \rangle . \tag{2.9}$$

With the n + 1 slices of the interval β, the end results are very similar to the propagation of a particle in Euclidean time.[6)]

Since we calculate the trace, there are now n + 1 variables of integration in Eq. (2.5) where no $\hbar$ appears now. In Eq. (2.6) a is identified as $\beta/(n + 1)$, and the additional variable to integrate is $Q_i = Q_f$.

We learn therefore that the thermodynamics of the quantum mechanical particle is equivalent to a periodic one-dimensional crystal, and the temperature T of the heat bath appears as $\beta = (n + 1)a$ where β is the length of the crystal. It is important to distinguish between the physical heat bath which is the environment of the quantum mechanical particle in the original thermal problem and the heat bath of the stochastic procedure in generating the Markov chain. The two have nothing to do with each other.

We apply now the above described machinery to the double well potential

$$V(Q) = -\frac{1}{2}\mu^2 Q^2 + \lambda Q^4 \quad , \tag{2.10}$$

with two minima as depicted in Fig. 2. This model was studied in detail by Creutz and Freedman.

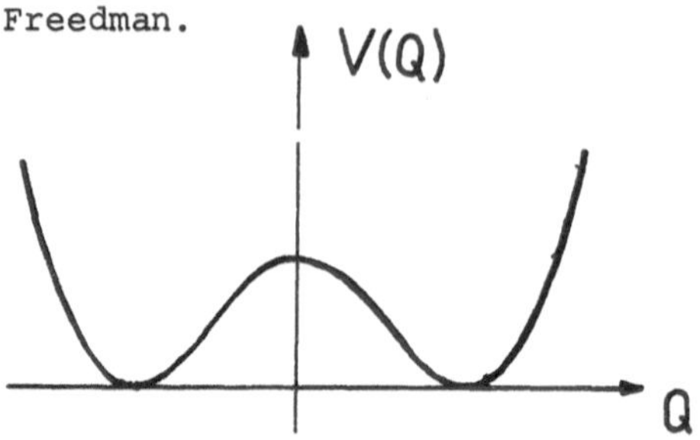

Fig. 2

When the temperature is low so that kT is much smaller than the ground state energy E_0, we can extract information about the ground state properties of the system.

Fig. 3 shows the probability distribution of the particle in the ground state. The two maxima of the probability distribution around the potential minima is nicely seen (the next three figures are results from Creutz and Freedman[5]). The stochastic points sit on the dashed line of the exact calculation.

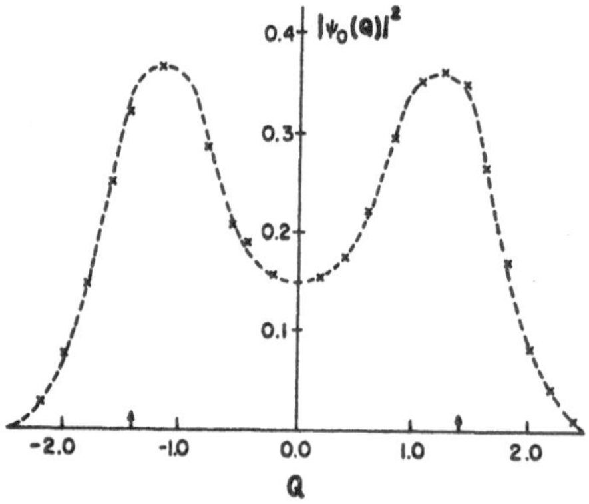

Fig. 3

We have a quantum mechanical tunneling problem here. The average value of the particle's position oscillates between the two potential minima for $kT \ll E_0$. This is shown in Fig. 4.

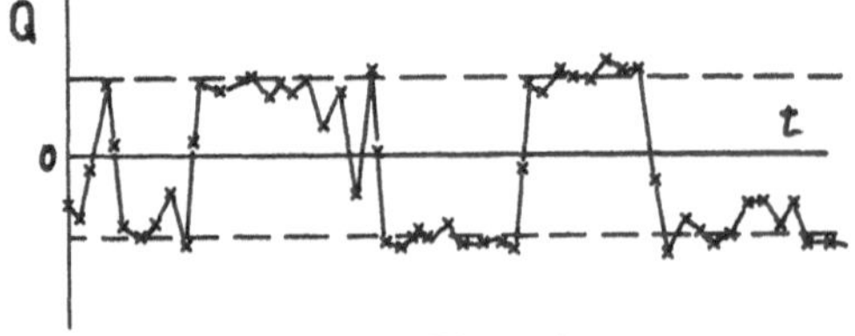

Fig. 4

With antiperiodic boundary condition, there is an instanton solution (the famous tanh function) to the classical equation of motion in Euclidean time. Creutz and Freedman find it on the one-dimensional crystal with the twisted boundary condition as depicted in Fig. 5.

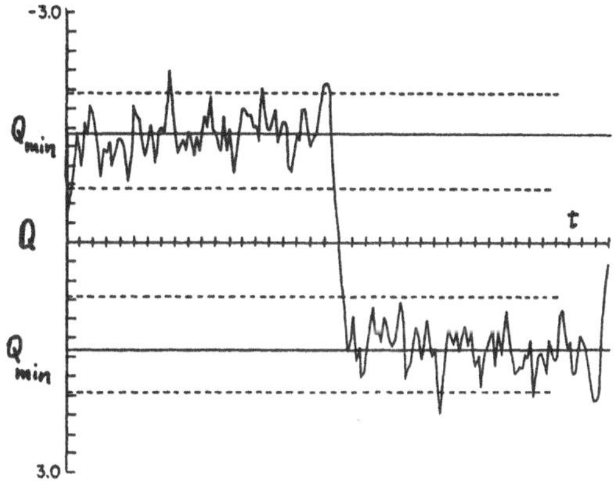

Fig. 5

3. The pure gauge sector of QCD

Since we gained some confidence about the applicability of the method, it is time to turn to Quantum Chromodynamics.[7] First I will discuss the pure gauge sector of the theory.

The spirit of the calculation is the same as in the simple quantum mechanical example. The functional integrals are four-dimensional now in Euclidean space-time. We take a finite volume (box) in this 4 - d space and introduce a lattice regularization along Wilson's suggestion. Gauge fields live on links in lattice QCD and quark fields live on lattice sites.[8] We have a system with a finite number of degrees of freedom and we have to repeat a similar calculational procedure as in the simple one-dimensional example. The cut-off of the regularized theory is the lattice spacing a.

3a. String tension

The potential energy v(r) between a heavy quark-antiquark pair at large separation r defines the string tension K in the pure gluon sector of the theory,

$$V(r) \sim -\frac{4}{3}\frac{\alpha(r)}{r} + Kr \tag{3.1}$$

Eq. (3.1) describes asymptotic freedom at short distances and the linear potential is associated with quark confinement. One often pictures the confining force as a chromoelectric flux tube between the quark-antiquark pair. K essentially tells you the energy per unit length stored in this flux tube.

K also relates phenomenologically to the slope $\alpha'(o)$ of Regge trajectories, if high angular momentum excitations are described as rotating strings:

$$K \simeq \frac{1}{2\pi\alpha'(0)} \quad . \tag{3.2}$$

Heavy particle spectroscopy, which is based on the potential (3.1), and universal Regge slopes tent to give a value for K somewhere around

$$\sqrt{K} \simeq 400 - 500 \text{ MeV}. \tag{3.3}$$

The string tension was the first physical quantity determined rather accurately by the above presented stochastic procedure. Since K is a dimensional physical quantity, it depends on the bare coupling constant g(a) as

$$K = \text{const}\cdot\frac{1}{a^2}\,e^{-\frac{2c}{g^2(a)}} \tag{3.4}$$

This singular behavior on the coupling constant is similar to the case of hadron masses as discussed in the introduction. The constant C is calculable in SU(3), and for the sake of simplicity I did not write out a factor in front of the exponent which changes slowly as a function of g^2. In the actual numerical calculation it is taken into account.

The results of Creutz[9)] are shown in Fig. 6 for pure SU(3) gauge theory. The straight line on the logarithmic plot is the envelope of the logarithm of properly chosen Wilson loop ratios to project out

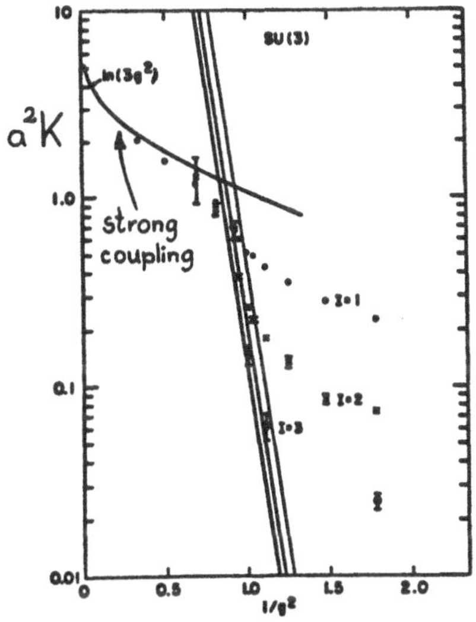

Fig. 6

the coefficient of the area term (a^2K) at a quark-antiquark separation I. This envelope is just a^2K for large Wilson loops, since everything is measured on the lattice in lattice spacing units.

For large bare coupling g we are in the strong coupling limit of the theory which is not physical. For very small g, the lattice spacing a becomes very small in physical units and therefore the total volume of the lattice becomes too small to pick up non-perturbative infrared effects. This is why the Wilson loops of fixed site in lattice spacing units break away from the straight line envelope.

As displayed in Fig. 6 there is only a narrow window of vulnerability where one can observe the right behavior of K as given by Eq. (3.4). In this narrow window the system follows the behavior as expected in the continuum limit $a \to 0$.

The combined results of two published calculations[9,10] yield a very important result

$$K^{\frac{1}{2}} = (5 \pm 2)\,\Lambda_{\overline{MS}} \qquad (3.5)$$

in SU(3) pure gauge theory. Clearly, on the lattice one can only relate K to the strength of the quark-gluon coupling in terms of $\Lambda_{\overline{MS}}$. To find this relation also required some important calculation[11] to connect the bare coupling constant g(a) to, say, traditional regularization schemes where $\Lambda_{\overline{MS}}$ is defined.

From Eqs. (3.3) and (3.5) we estimate $\Lambda_{\overline{MS}}$ to be somewhere around 100 MeV. If this result will not change much in the presence of quark vacuum polarization, it may have interesting consequences for the proton decay in SU(5) unification.

3b. Critical temperature of thermal quark liberation

There has been a long-standing conjecture that a phase transition will take place between the low temperature confining phase of QCD and the deconfining gluon plasma phase. Quarks can move freely in the hot gluon plasma.

Recent calculations[12] give strong numerical evidence that the quark liberating phase transition occurs, indeed, in the gauge sector of QCD. The calculation goes as follows.

We insert a heavy quark-antiquark pair in the vacuum at large separation. At zero temperature some chromoelectric flux develops between the quark and antiquark and we experience a confining force law. Now we heat the system by repeating the calculation of the force at finite temperature. We saw in the case of the simple quantum mechanical example how to do the calculation at finite physical temperature.

One finds a critical temperature T_c where the confining force between the heavy quark and antiquark suddenly disappears. This phase transition is a bulk property of the system and therefore easy to detect in the calculation.

What happens somehow is that the vacuum as a medium which supported the confining chromoelectric flux gradually fills up with glueballs as the temperature is rising, and at the critical temperature T_c suddenly some gluon plasma forms. Quarks can move freely in this hot gluon plasma.

Some recent results[13] on the quark liberating phase transition are shown in Fig. 7 for SU(3). The order parameter

$$W = \exp(-\beta F) \quad , \tag{3.6}$$

where F is the free energy of an isolated heavy quark in the heat bath, is plotted against the temperature T ($\beta = 1/kT$) in some dimensionless units. We observe an abrupt change in the order parameter at the

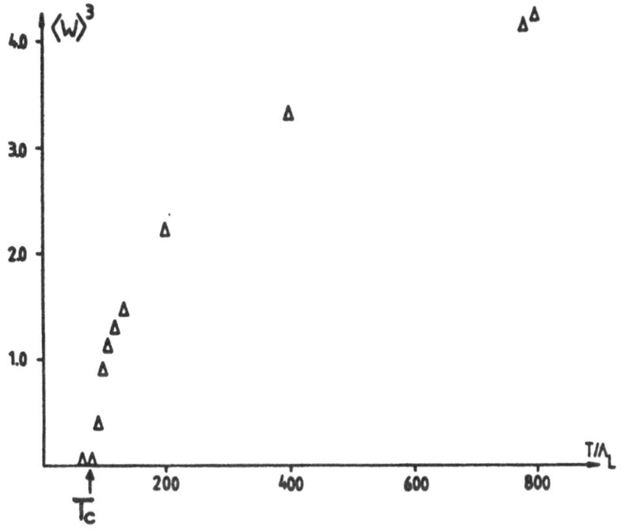

Fig. 7

critical temperature T_c.

Below T_c the field around the isolated quark forms a collimated chromoelectric flux. Above T_c we find a Debye screened Coulomb field as shown in Fig. 8.

Fig. 8

The critical temperature T_c was found to be

$$T_c \simeq \frac{\sqrt{K}}{2} \tag{3.7}$$

in the SU(3) model. The numerical value of T_c is somewhere around 200 MeV. The high energy heavy ion physics community regards the phase transition of thermal quark liberation and the low value of T_c as a real challenge.

3c. Mass gap and glueball excitations

It would be very interesting to determine numerically the mass spectrum of the lowest glueball states. Previous attempts[14)] were based on the study of the exponential decay of correlation functions, like

$$\langle 0 | G^2_{\mu\nu}(x)\, G^2_{\mu\nu}(0) | 0 \rangle \quad , \tag{3.8}$$

where $G_{\mu\nu}$ designates the field strength of the gluon field. The exponential decay of the correlation function at large separation is governed by the lowest glueball state which can be excited from the vacuum by the composite operator $G^2_{\mu\nu}$.

Recently, Wilson proposed a new improved scheme[15)] which allows the extraction of glueball masses from correlation functions of composite operators at short distances.

The first result in the SU(3) model was recently reported[16)] for the mass of the 0^+ glueball to be

$$m(0^+) = (2.5 \pm 0.8)\sqrt{K} \quad . \tag{3.9}$$

The main source of the thirty percent error in Eq. (3.9) is the uncertainty in the relation between K and $\Lambda_{\overline{MS}}$ (see, Eq. (3.5)). The numerical value of the 0^+ glueball is somewhere around 1.0-1.2 GeV.

It appears that the ratio of the critical temperature T_c and the glueball mass $m(0^+)$,

$$\frac{T_c}{m(0^+)} = 0.19 \pm 0.02 \quad , \tag{3.10}$$

is very accurately known now in the SU(3) gauge model.

3d. Gluon condensate in the vacuum

The vacuum state of Quantum Chromodynamics, even in the absence of quark vacuum polarization, describes a complicated medium whose physical properties are only poorly known.

Unfortunately, we have learned only very little so far about the detailed structure of the vacuum from the numerical investigations.

We know that the vacuum as a medium is responsible for confinement, and when heated it suffers a deconfining phase transition at a critical temperature. Nothing is known about the nature of the gluon condensate which is responsible for those phenomena.

We cry for the kind of simple pictures which describe so beautifully some of the simple spin models. Fig. 9 shows the two-dimensional x-y model as it evolves and equilibrates from some configuration at t = 0 to a low temperature state in 1000 Monte Carlo sweeps through the lattice.[17] One can clearly see the Kosterlitz-Thouless vortices as they come alive from the given initial configuration when the system equilibrates in search of the important spin configurations. Of course, we know the important role of the Kosterlitz-Thouless vortices

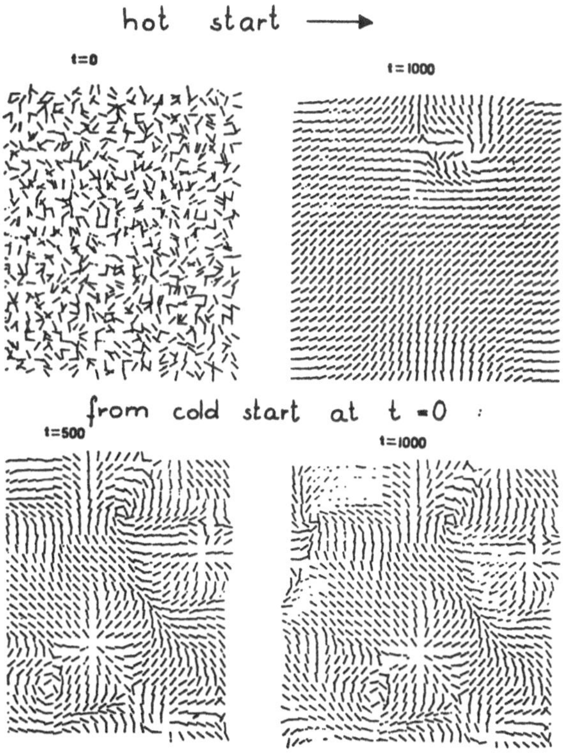

Fig. 9

in the 2-d x-y model. The upper part of Fig. 9 describes how the system equilibrates from a hot start to some low temperature below T_c. The lower part shows the more rapid evolution from a cold start to the same equilibrium temperature.

It would be very satisfying to see that sort of qualitative insight into the vacuum structure of QCD.

Recently we have witnessed some efforts to establish the presence of some non-perturbative gluon condensate in the QCD ground state.[18) The quantity

$$G = \left\langle \frac{\alpha}{\pi} \sum_{\mu\nu,i} G^i_{\mu\nu} G^i_{\mu\nu} \right\rangle \quad , \quad \alpha = \frac{g^2}{2\pi} \tag{3.11}$$

was extracted from Monte Carlo data for SU(2) lattice gauge theory. In Eq. (3.11) G is the vacuum expectation value of the square of the field strength tensor $G^i_{\mu\nu}$ in Euclidean space-time metric where a summation is understood over μ,ν and the color index i.

The trace anomaly equation for the energy-momentum tensor connects G to the vacuum energy.[19) G is of course ultraviolet divergent with a leading term $\sim a^{-4}$ as a function of the cutoff a. To interpret G physically, the authors of Ref. 18 try to subtract from Eq. (3.11) the quantity G_{pert} which is defined as the perturbative expansion of Eq. (3.11) into a power series as a function of g^2. The conjecture is that $G-G_{pert}$ scales under the renormalization group as an operator of dimension [mass^4],

$$G - G_{pert} \sim \frac{1}{a^4} e^{-\frac{4c}{g^2}} \tag{3.12}$$

The subtracted expression $G-G_{pert}$ should describe then the energy decrease of the ground state in the presence of some non-perturbative gluon condensate, relative to the perturbative vacuum.

The numerical results of Monte Carlo calculations to find Eq. (3.12) as a function of g^2 are, I believe, against the above naive expectation. Two non-trivial terms are known analytically in the expansion of G_{pert}. If one includes the third term in the expansion, its unknown coefficient can be obtained from the Monte Carlo results for G in the very small g^2 region, where $G \approx G_{pert}$ to a great accuracy, according to Eq. (3.12). The difference $G-G_{pert}$, so obtained, does not scale according to Eq. (3.12) in the region $2.2 < 4/g^2 < 2.8$ which is the narrow window of vulnerability.[20) The coefficient in the exponent is smaller by about a factor of two.

One can, of course, insist upon a fit of the form

$$G = G_{pert} + \frac{1}{a^4} e^{-\frac{4c}{g^2}} \tag{3.13}$$

to the numerical results on G by keeping many higher terms in G_{pert} with alternating signs. They cancel for very small g^2 and $G-G_{pert}$ looks good in the window. However, G_{pert} in this window does not look like a perturbation expansion anymore. It builds up from large oscillating terms.

Clearly, one should do something better. It is difficult, if not impossible, to probe a condensate with the above method, when the natural length scale of the structure in the condensate should be of the order of 1 fermi, and not the cut-off a. Quite correspondingly, I think, it would be very difficult to see the Kosterlitz-Thouless vortices in Fig. 9 from local measurements.

4. Stochastic treatment of the fermion problem

During the last twelve months we have witnessed considerable effort to develop Monte Carlo methods for the numerical study of quantum systems with fermionic degrees of freedom. This outstanding problem is of great importance for applications in quantum field theories, condensed matter physics, and nuclear physics

Previous techniques[21)] were slow when a very large number of fermionic degrees of freedom were involved, since the computational time required for a Monte Carlo sweep through the lattice was always proportional to the square of the crystal volume, or even worse.

Recently, this problem was solved by Hirsch, Scalapino, Sugar and Blankenbeckler in one space and one time dimension.[22)] They follow the evolution of fermion world lines along the Euclidean time direction with an update time independent of the lattice volume. The method is fast and efficient in applicaitons. The generalization of this ingenious idea to higher dimensions is desirable.

I will follow here the more standard strategy and work directly with a new effective action of the boson fields when the fermionic degrees of freedom are integrated out. Though the effective action becomes non-local in the presence of the fermion determinant, a new procedure seems to maintain the efficiency of the standard Monte Carlo technique in the sense that the update time on a site is independent of the lattice volume.[23)] The method is applicable in any number of dimensions.

4a. Fermion determinant

For a general presentation, I will consider now the Euclidean action

$$S = S_0(U) + \sum_{ij} \bar{\psi}_i M_{ij}(U) \psi_j \tag{4.1}$$

in four dimensions. It describes the interaction of a boson field U_i with a fermion field ψ_i, and the subscripts on the fields refer to the lattice points. Spin and internal symmetry indices are supressed, for simplicity. The matrix $M_{ij}(U)$ designates both kinetic and mass terms for the fermion field, and couplings to the boson field. $S_0(U)$ describes the pure bosonic part of the Euclidean action.

It is important to note that most of the interesting models in quantum field theory, condensed matter physics, and nuclear physics can be brought to a bilinear form in the fermion fields.

The fermion Green's functions can be calculated by inserting sources into the path integral

$$Z(\bar{\eta},\eta) = \int D\bar{\psi} D\psi DU \; \exp\left[-S + \sum_i (\bar{\eta}_i \psi_i + \bar{\psi}_i \eta_i)\right] \quad . \tag{4.2}$$

By taking the functional derivatives and integrating out the Grassman variables, the fermion correlation function can be written as

$$\begin{aligned} \langle \bar{\psi}_i \psi_j \rangle &= \frac{\delta^2}{\delta\eta_i \delta\bar{\eta}_j} \ln Z(\bar{\eta},\eta) \Big|_{\bar{\eta}=\eta=0} \\ &= \frac{1}{Z} \int DU \, M^{-1}_{ij}(U) \exp\left[-S_{eff}(U)\right] \quad , \end{aligned} \tag{4.3}$$

where Z is the partition function (normalization integral) of the boson-fermion system.

The effective action is given by

$$\exp\left[-S_{eff}(U)\right] = \det\left[M(U)\right] \exp\left[-S_0(U)\right] \quad , \tag{4.4}$$

and I assume, for simplicity only, that the fermion determinant has positive sign.

I apply now the Metropolis Monte Carlo method to the evaluation of the functional integral in Eq. (4.3). Other Euclidean Green's functions can be treated similarly.

It was shown by Scalapino and Sugar[21)] that a local change $U \rightarrow U + \delta U$ implies

$$\frac{\exp\left[-S_{eff}(U+\delta U)\right]}{\exp\left[-S_{eff}(U)\right]} = \det\left[1 + M^{-1}(U)\delta M(U)\right] \cdot \frac{\exp\left[-S_0(U+\delta U)\right]}{\exp\left[-S_0(U)\right]} \tag{4.5}$$

With local boson-fermion coupling the non-trivial change δM in the fermion matrix is restricted to the neighborhood of the updated lattice site. Consequently, we need only a few inverse elements of the large matrix M in each Metropolis step.

At that point I depart from standard procedures. Since the results of a Monte Carlo calculation are always subject to some statistical inaccuracy, it is reasonable to evaluate the decision-making step stochastically.

I will calculate the inverse matrix elements of M by some modification of a stochastic method which was first suggested by J. von Neumann and S. M. Ulam, but never published by them. It is a very efficient method for the approximate summation of the Neumann series defined by the inverse of the operator M.

Assume that the inverse of a matrix M of order m is desired and let H = I - M, where I is the unit matrix. For the method to be applicable, it is necessary and sufficient that the eigenvalues of the matrix H_{ij} are less than one in absolute value.

Note that the above condition can always be arranged by proper normalization. The matrix elements $(M^{-1})_{ij}$ are given by the solutions of the linear system of equations Mx = b, with unit driving vectors on the right-hand side. This equation is equivalent to $(2/\mu)M^+Mx = (2/\mu)M^+b$ where μ is the first norm of the matrix M^+M. The driving vector $(2/\mu)M^+b$ may be decomposed into a linear combination of unit vectors and, with the replacement $M \to (2/\mu)M^+M$, the method applies even in the worst case.

I decompose the matrix element H_{ik} into $H_{ik} = P_{ik} \cdot R_{ik}$ with the restriction that $P_{ik} > 0$ and $\sum_{r=1}^{m} P_{ir} < 1$ for all values of i. Consider a random walk on the domain of integers 1,2, . . . ,m. The walk begins at some selected point i and proceeds from point to point with the transition probabilities P_{ik}. The walk stops after k steps at some point s_k with the stop probability $P_{s_k} = 1 - \sum_{r=1}^{m} P_{s_k r}$.

When the walk stops, a score S_{ij} is registered for the elements in the ith row of the inverse matrix. It is defined by the product of the residues $R_{s_r s_{r+1}}$ along the trajectory $i \to s_1 \to s_2 \to . . . \to s_k = j$ divided by the stop probability P_j:

$$S_{ij} = \begin{cases} 0 & \text{if } s_k \neq j \\ R_{is_1} R_{s_1 s_2} \ldots R_{s_{k-1} j} P_j^{-1} & \text{if } s_k = j \end{cases} \tag{4.6}$$

I will prove that the expectation value of the random variable S_{ij} is $(M^{-1})_{ij}$. Indeed, the probability of a walk to follow some trajectory $i \to j$ and to stop at j is $P(i \to j) \cdot P_j = P_{is_1} P_{s_1 s_2} \ldots P_{s_{k-1} j} P_j$. The expected score is given by the sum over all trajectories form i to j:

$$\langle S_{ij} \rangle = \sum_{i \to j} P(i \to j) P_j S_{ij} = \sum_{i \to j} P(i \to j) R(i \to j) \tag{4.7}$$

where $R(i \to j)$ is the product of the residues along the trajectory. Since $P_{ij} \cdot R_{ij} = H_{ij}$, Eq. (4.7) is recognized as the Neumann series expansion for $M^{-1} = (1 - H)^{-1}$. The term δ_{ij} in the Neumann series is generated by walks which stop immediately.

It is easy to prove that the variance σ_{ij}^2 of the random variable S_{ij} is given by

$$\sigma_{ij}^2 = (Q^{-1})_{ij} P_j^{-1} - (M^{-1})_{ij}^2$$

where $Q = (I - K)^{-1}$ with $K_{ij} = H_{ij} \cdot R_{ij}$. The variance of S_{ij} is finite, provided the Neumann series for $Q = (I - K)^{-1}$ exists.

The statistical error on $(M^{-1})_{ij}$ is given by $\sigma_{ij}/\sqrt{N}$ for N walks which all begin at point i. For a given statistical accuracy in the decision-making step of the Metropolis procedure, the required number of walks does not depend on the size of the matrix. Therefore, the update time in this stochastic procedure is independent of the lattice volume. My tests involved matrices of the order of 10^4, or larger.

Before we test the method on some four-dimensional fermion problem, it is appropriate to discuss a radical approximation to the fermion Monte Carlo procedure, since it was successfully applied to Quantum Chromodynamics.

4b. Hadron spectrum without internal quark loops

Recently, some very interesting results were obtained in QCD concerning chiral symmetry breaking and the spectrum of light hadrons. In the so-called quenched approximation[24)] the fermion determinant (det M) was set to one during the Monte Carlo procedure. That approximation corresponds to neglecting internal quark loops in Feynman diagrams. The hope is then that quark vacuum polarization will not play an important role in the mechanism of chiral symmetry breaking, or in the spectrum of light hadrons.

When I first studied the papers from Brookhaven[24)] I was somewhat sceptic about some of the quantitative results. Since then Claudio Rebbi kindly called my attention to some tricks they used in the numerical analysis.[25)] I assume now that the published numbers are correct within the limits of the quenched approximation.

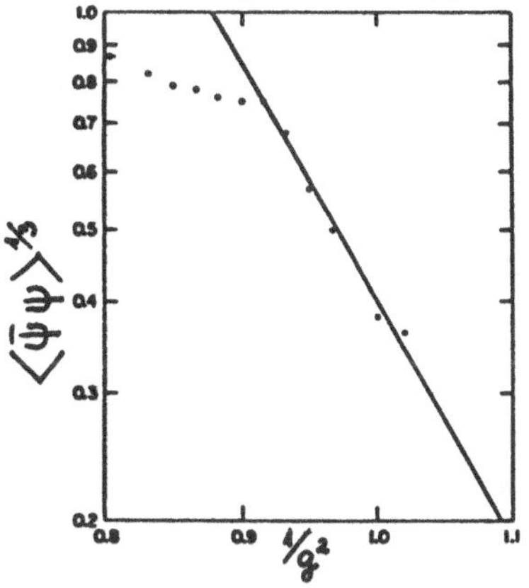

Fig. 10

Rebbi and his co-workers find spontaneous chiral symmetry breaking in the SU(2) model with Susskind fermions. The symmetry breaking is also seen in the SU(3) model with naive lattice fermions (first paper in Ref. 24). Fig. 10 shows the non-vanishing expectation value of $\bar{\psi}\psi$ as a function of $1/g^2$ in the SU(3) model with naive lattice fermions, in the limit when the bare quark masses vanish. Quark vacuum polarization, as we said, is neglected in the calculation.

Fig. 11 shows from the first paper of Ref. 24 the light hadron spectrum as a function of the bare quark mass m_q in the SU(3) model with Wilson fermions. Very good results are reported in the absence of quark

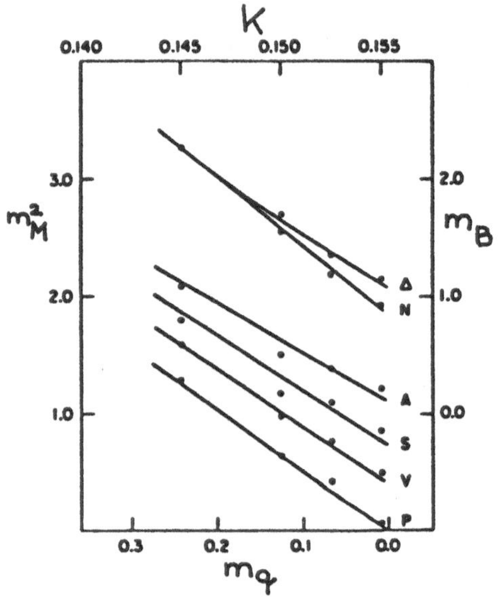

Fig. 11

vacuum polarization, but for further details I have to refer you to the original papers.[24)]

Perhaps my only somewhat sceptic remark about Fig. 11 is the very modest lattice size of $6^3 \cdot 10$ sites. The small lattice may distort to some extent the physics we want to study. I assume that results on larger lattices with accurate points will be available soon in the quenched approximation.

The effects of quark vacuum polarization remain to be seen.

4c. Method to treat the complete fermionic theory

After this excursion where det M was set to one, we return now to the complete fermion problem.[26)]

I will modify now the Neumann-Ulam algorithm of Section 4a for better efficiency in fermionic Monte Carlo procedures. It is easy to realize that during a walk which started at point i, one can register the product of residues at each pass through the point j. I define a new random variable $\tilde{S}_{ij}$ as the sum of the products of residues, adding a new term to the score at each pass through the point j. The stop probabilities are eliminated from the random variable $\tilde{S}_{ij}$, but they still govern the average length of a walk.

It is straightforward to show that $\langle S_{ij} \rangle = \langle \tilde{S}_{ij} \rangle$ when the stop probability P_j is positive. I have also proved that the expectation value $\langle \tilde{S}_{ij} \rangle$ is equal to $(M^{-1})_{ij}$ when the stop probability P_j vanishes. The original method does not apply in this case.

In order to compare efficiencies, I choose a simple case when all $R_{ij} = 1$ and P_j is positive. A necessary and sufficient condition for the variance of the random variable $\tilde{S}_{ij}$ to be smaller than σ_{ij}^2 is $P_j < e_j/(2 - e_j)$, where e_j designates the escape probability from the point j. In practice, this condition is enforced by the nature of the fermion problem, and the modified method is much more efficient.

In the special case when all stop probabilities vanish, one has to stop by fiat. Some bias is introduced then, since the Neumann series is truncated after a finite number of terms.

The modification described above is probably known to some experts on stochastic methods and the special case when the random walk is stopped by fiat appears in the Russian literature.

I tested my stochastic fermion method on a four-dimensional boson-fermion model which was first suggested by Scalapino and Sugar.[21)] The fermion matrix M in Eq. (4.1) is specified now as

$$M_{ij} = -\Delta_{ij} + (m^2 + gU_i^2)\delta_{ij} \tag{4.8}$$

where U_i and ψ_i are a scalar boson field and a spinless fermion field, respectively. Δ_{ij} defines the Laplacian operator on the lattice, m is the bare fermion mass in lattice spacing units, and g designates the dimensional boson-fermion coupling constant.

The functional integral is calculable analytically in this model, and one finds

$$D(i-j) = \langle \bar{\psi}_i \psi_j \rangle = \left(-\Delta + m^2 + \tfrac{1}{2} g\right)_{ij}^{-1} \tag{4.9}$$

The fermion-boson interaction generates a mass term dynamically, and the renormalized fermion mass is given by $m_r = (m^2 + 1/2g)^{1/2}$.

Some results of my calculations are presented in Fig. 12. The complete fermion mass was generated dynamically with the choice $m = 0$. $D(i - j)$ is depicted for a lattice of $8^3 * 16$ sites with periodic boundary condition and coupling constant $g = 2.6a^{-2}$ (in the actual calculations the lattice spacing a was set to unity). The free fermion propagator, with renormalized mass $m_r^2 = 1.3a^{-2}$ on the same lattice size, is represented by the solid line. The agreement of the numerical points with the exact analytic form is very satisfactory (the statistical errors are practically not visible on the logarithimic plot). Results are also presented for a lattice of 10^4 sites with $m_r^2 = 0.25a^{-2}$.

The quenched approximation[24)], where one neglects the fermionic vacuum polarization effects from the fermion determinant in the effective action, is also presented in Fig. 12. The dashed line is the fit of a free fermion propagator of mass $m_q^2 = 1.02a^{-2}$ to the results of the quenched approximation on the $8^3 * 16$ periodic lattice with $g = 2.6a^{-2}$ and $m = 0$. The contribution of the fermion loops is clearly seen and accurately calculated: $m_q^2/m_r^2 = 0.78$. In Quantum Chromodynamics I expect this ratio to be smaller.

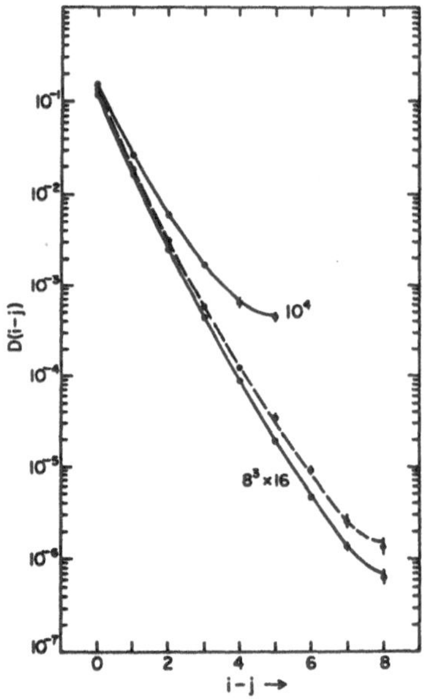

Fig. 12

The speed and efficiency of my stochastic method is very promising. It took only about 3 hours on a VAX 11/780 to calculate the complete fermion propagator on the $8^3 * 16$ lattice, with relative errors which are only a few percent even for a separation of eight links along the fourth direction.

I hope that I can make some progress in the near future on the application of this stochastic fermion method to the numerical study of Quantum Chromodynamics.

REFERENCES

1. For a review, see: C. Itzykson and J. B. Zuber, Quantum Field Theory, McBraw-Hill Book Company (1980).

2. D. J. Gross and F. Wilczek, Phys. Rev. Lett., 30, 134 (1973); H. D. Politzer, Phys. Rev. Lett., 30, 1346 (1973).

3. W. Celmaster and R. J. Gonsalves, Phys. Rev. D20, 1420 (1979).

4. For a review, see: J. Ellis, TH-2174-CERN preprint (1981).

5. For a review, see: M. Creutz and B. Freedman, BNL 28588 Brookhaven preprint (1980).

6. R. P. Feynman: Statistical Mechanics, W. A. Benjamin, Inc. (1972).

7. The application of the stochastic method for the numerical study of QCD was suggested by K. G. Wilson in 1978.

8. K. G. Wilson, Phys. Rev. D10, 2445 (1974).

9. The pioneering SU(2) results are found in M. Creutz, Phys. Rev. D21, 2308 (1980); for an important study of spin systems, see: M. Creutz, L. Jacobs and C. Rebbi, Phys. Rev. Lett. 42, 1390 (1979); for the first results on the SU(3) string tension, see: M. Creutz, Phys. Rev. Lett. 45, 313 (1981).

10. E. Pietarinen, Nucl. Phys. B190 [FS3], 349 (1981).

11. A. Hasenfratz and P. Hasenfratz, Phys. Lett 93B, 165 (1980).

12. This phase transition and the critical temperature T_c was found first by L. D. McLerran and B. Svetitsky, Phys. Lett. 98B, 195 (1981); J. Kuti, J. Polónyi and K. Szlachanyi, Phys. Lett. 98B, 199 (1981); later it was further confirmed and substantiated by K. Kajantie, C. Montonen and E. Pietarinen, Z. Physik C9, 253 (1981); J. Engels, F. Karsch, I. Montvay, H. Satz, Phys. Lett. 101B, 89 (1981).

13. I. Montvay and E. Pietarinen, DESY 82-077 preprint (1981).

14. B. Berg, Phys. Lett. 97B, 401 (1980); G. Bahnot and C. Rebbi, Nucl Phys. B180 [FS2], 469 (1980).

15. K. G. Wilson, talk given at the Abingdon meeting on Lattice Gauge Theories (1981).

16. B. Berg and A. Billaire, TH-3230-CERN preprint (1982).

17. C. Kawabata and K. Binder, Solid State Comm. 22, 705 (1977).

18. A. DiGiacomo, G. C. Rossi, Phys. Lett. 100B, 481 (1981); T. Banks, R. Horsley, H. R. Rubinstein and V. Wolf, WIS-81/8 Weizmann Institute preprint (1981).

19. M. A. Shifman, A. I. Vainshtein, V. I. Zakharov, Nucl. Phys. B165, 45 (1981).

20. R. Brower, J. Kuti, M. Nauenberg, J. Polónyi, preprint in preparation.

21. F. Fucito, E. Marinari, G. Parisi and C. Rebbi, Nucl. Phys. B180 [FS2], 369 (1981); D. J. Scalapino and R. L. Sugar, Phys. Rev. Lett. 46, 519 (1981); D. Weingarten and D. Petcher, Phys. Lett. 99B, 333 (1981); R. Blankenbecler, D. J. Scalapino and R. L. Sugar, Phys. Rev. C24, 2278 (1981); D. J. Scalapino and R. L. Sugar, Phys. Rev. B24, 4295 (1981); E. Marinari, G. Parisi and C. Rebbi, Nucl Phys. B190 [FS3], 734 (1981); H. Hamber, Phys. Rev. D24, 951 (1981); A. Duncan and M. Furman, Nucl. Phys. B190 [FS3], 767 (1981); Y. Cohen, S. Elitzur and E. Rabinovici, Hebrew University preprint, 1981; the use of the Wilson hopping parameter K in the fermionic Monte Carlo procedure for the acceleration of the calculation was first suggesteg by I. O. Stamatescu, MPI-PAE/PTh 60/80 Max Planck Institute preprint (1980); the hopping parameter expansion in the Monte Carlo procedure was first reported by A. Hasenfratz and P. Hasenfratz, Phys. Lett. 104B, 489 (1981).

22. J. E. Hirsch, D. J. Scalapino, R. L. Sugar and R. Blankenbecler, Phys. Rev. Lett. 47, 1628 (1981).

23. For the description of the method, and for further references, see: J. Kuti, NSF-ITP-81-151 Santa Barbara preprint (1981).

24. H. Hamber and G. Parisi, Phys. Rev. Lett. 47, 1792 (1981); E. Marinari, G. Parisi and C. Rebbi, 47, 1795 (1981); D. Weingarten, Indiana University preprint, 1981.

25. I appreciate a very useful discussion with Claudio Rebbi on that matter.

26. During the preparation of the notes a very interesting preprint reached me in which the authors apply the hopping parameter expansio to the fermion Monte Carlo problem in QCD using fairly large lattice volumes: A. Hasenfratz, Z. Kunszt, P. Hasenfratz and C. B. Lang, TH-3220-CERN preprint (1981).

SPONTANEOUSLY BROKEN AND DYNAMICALLY ENHANCED GLOBAL AND LOCAL SYMMETRIES

Jürg Fröhlich
Institut des Hautes Etudes Scientifiques
35, route de Chartres
F-91440 Bures-sur-Yvette

Abstract.

I re-examine the notions of spontaneously broken, global and local symmetries and discuss them in terms of some examples in quantum field theory or statistical mechanics. I then briefly recall some basic ideas and facts about the renormalization group. They are used to introduce and discuss the concept of dynamically enhanced (or "generated") asymptotic symmetries.

Contents.

Remark. Most of the ideas and results described in these notes have been worked out in collaboration with T. Spencer to whom I am deeply indebted for his clear insights and his generosity. I have also benefitted from collaborations with D. Brydges, G. Morchio, C. Pfister, E. Seiler, B. Simon and F. Strocchi to whom I extend my gratitude.

These notes are intended for light reading. The interested reader is advised to consult the following references for further study :

Sect. 1 : [1,2,3,4] .
Sect. 2 : [5,6,7] and the reviews [2,3,4,8,9] .
Sect. 3 : [2,10,11] and in particular [12] .
Sect. 4 : [13]
Sect. 5 : [14,15,16] (see also the contribution of H.P. Dürr).

1. Remarks on the historical development of symmetry concepts; basic definitions.

1.1.

We distinguish two aspects of symmetry

a) a geometrical, static aspect, and
b) a dynamical aspect.

Moreover, dynamical symmetries can either be

b1) global symmetries, or
b2) local "symmetries".

a) The physics of the antique and of the middle ages (Ptolemy, Kepler...), before Galilei and Newton, only knew the static, geometrical aspect of the concept of symmetries. (The orbits of the planets were believed to have high symmetry. An attempt was made by Kepler to explain the interplanetary distances in terms of the platonic solids. Matter was conceived as being built of highly symmetric "elementary bodies", etc.).

The geometrical aspect of symmetry is of course still extremely important in molecular physics, crystallography, condensed matter physics, biophysics. Historically, it has been an important root in the development of group theory. In a more algebraic outfit, it still appears in every branch of physics in discussions of the symmetries of invariant states (vacuum - or equilibrium states) of physical systems.

b) The idea that it is not the orbits of physical systems which necessarily exhibit high symmetries, but that it is the dynamical laws of physical systems which may admit invariance or symmetry groups could of course appear only after dynamics was introduced into physics, i.e. after the discovery of Newtonian mechanics.

I now briefly characterize static and global dynamical symmetries of physical systems abstractly : We consider a physical system , S . The family of all possible states of S is called X , its elements $x,y,\ldots$. We assume that X carries the action of a group G , i.e. with each $x \in X$ and each $g \in G$ we associate a state $x_g \in X$, the image of x under the transformation g .

a) A state x of a physical system has a static symmetry, described by a subgroup $H \subseteq G$, if

$$x_h = x \text{ , for all } h \in H .$$

(H is the stability group of x . All states on the same G-orbit have conjugate stability groups). An orbit of a dynamical, physical system, S , is a mapping from the real line, the time axis, into the space of states, X , of S , i.e.

$$\mathbb{R} \ni t \longmapsto x(t) \in X .$$

The family of all possible orbits is denoted O .

b1) S has a global (dynamical) invariance - or symmetry group $H \subseteq G$ if with each $x(\cdot) \in O$ the trajectory

$$\{x(t)_h : t \in \mathbb{R}\}$$

is again an orbit of S , i.e. an element of O , for all $h \in H$. This means that

$$x(t)_h = x_h(t) ,$$

(where $x = x(0)$, x_h = image of x under h , $x_h(t)$ = orbit with initial condition x_h) .

b2) Next, we give a somewhat misleading, abstract characterization of local (dynamical) symmetries : Suppose S consists of (spatially separated, but coupled) constituents, $S_1,\ldots,S_n$, $n = 2,3,\ldots,$ i.e. $S \cong S_1 \times S_2 \times \cdots \times S_n$.

An orbit, $x(t)$, of S then consists of an n-tuple $(x_1(t),\ldots,x_n(t))$, where $x_i(t)$ is an orbit of the constituent system S_i , $i = 1,\ldots,n$. A subgroup $H \subseteq G$ is a "local symmetry group" of S if, for arbitrary $h_1,\ldots,h_n$ in H and an arbitrary orbit, $x(t)$, of S ,

$$x(t)_{\underset{\sim}{h}} \equiv (x_1(t)_{h_1},\ldots,x_n(t)_{h_n}) ,$$

$\underset{\sim}{h} = (h_1,\ldots,h_n)$, is again a possible orbit of S . Usually, it is not assumed that $\underset{\sim}{h}$ be time-independent.

"Local symmetries" have little in common with global symmetries, since for physical systems admitting a local symmetry group (an invariance group of gauge transformations of the second kind) one cannot devise any experiment which would distinguish between the two orbits $x(t)$ and $x(t)_{\underset{\sim}{h}}$. Thus, $x(t)$ and $x(t)_{\underset{\sim}{h}}$ really correspond to the same, physical orbit of S, expressed in different "internal coordinates".

The idea of "local symmetries" has led to the development of gauge theories, (H. Weyl, O. Klein, W. Pauli, Yang and Mills), in physics, and in mathematics it gave rise to the theory of fibre - and principal bundles.

The present trend in particle physics is to eliminate global symmetries from the fundamental dynamical laws, but to find dynamical laws admitting a (usually non-abelian) "local symmetry group", although the spatial symmetries (Poincaré covariance) are still global symmetries, unless gravity is included. Such dynamical laws are expressed in the form of gauge theories. That trend poses the interesting, theoretical problem of deriving the global, internal symmetries of phenomenological models or macroscopic descriptions as dynamically generated, asymptotic symmetries; (see Sect. 5).

1.2.

We now turn to the discussion of symmetry breaking. We start with the breaking of global symmetries : We distinguish between

i) Explicit breaking.
ii) Spontaneous breaking.
iii) Dynamical breaking.

We speak of an explicitly broken symmetry if the dynamics of a physical system contains a "small" term which is not invariant under a symmetry group leaving invariant the other terms of the dynamics. This concept is of importance when one tries to understand small masses (like the pion mass). It often arises in quantum field theories which, in the classical limit (tree approximation), have a symmetry that does not survive on the quantum level, because of anomalies. (Examples may be chiral symmetries (PCAC), dilation or conformal invariance). A different example for i) is the eightfold way. This topic is not discussed in my notes, although it is very important.

We say that a global symmetry group, H , of a physical system, S , is broken spontaneously (or dynamically) if H is an invariance group of the dynamics, but it is impossible to transform an orbit, $x(t)$, of S into the orbit $x(t)_h$, for some $h \in H$, by a sequence of local operations (local = local in space) without passing through states (or "configurations") of the system which have infinite energy, (or infinite action).

By considering the example of an infinite, three-dimensional ferromagnet one convinces oneself that the definition sketched above is appropriate.

Spontaneously broken, global symmetries can in general be characterized by a local order parameter, (in the example of the ferromagnet by the spontaneous magnetization = expectation value of the spin observable associated with a point or some microscopically small region).

It is not easy to distinguish between the concepts of spontaneous and dynamical symmetry breaking on an abstract level. (One might say that dynamical symmetry breaking does not manifest itself on the tree level and is non-perturbative, while spontaneous breaking appears already on the tree level and may be discussed perturbatively).

It is much more difficult to give an abstract characterization of the "spontaneous or dynamical breaking of local symmetries". This notion cannot mean that the physics of a system admitting a local symmetry group, i.e. a general covariance under changes of the internal coordinates, depends on the choice of local, internal coordinates. Broken, local symmetries cannot be characterized by (gauge-invariant) local order parameters.

The concept of a broken, local symmetry is intrinsically dynamical and must, in the author's opinion, be discussed in the context of a specific formulation of specific theories. One might vaguely describe it as follows : A system with a local, internal symmetry group H (assumed to be a compact Lie group) has always dynamical degrees of freedom carried by gauge fields which are indexed by generators of the Lie algebra of H . One might say that H is "broken spontaneously (or dynamically)" down to a subgroup $H_1 \subset H$ if the degrees of freedom associated with the coset space H/H_1 are "frozen out at small energies", i.e. are invisible at large distances. (But see Sect. 3).

A discussion of the concept of symmetry enhancement is postponed to Sects. 4 and 5.

2. Spontaneous breaking of global symmetries.

General references are [2,3,4,5,6]. General, group theoretical discussions of the possible symmetry breaking patterns in physical systems can be found e.g. in [17]. The breaking of spatial symmetries is a general phenomenon in the physics of matter at positive density (and temperature), in the thermodynamic limit : The boosts are always broken, rotation invariance is often broken (directional long range order; e.g. in liquid crystals), translation invariance is broken in systems with a crystalline equilibrium state (translational long range order). The mathematical understanding of systems with broken rotation - or translation invariance is however rudimentary.

There is a prominent example of the breaking of Lorentz invariance in a system at zero temperature and density : The boost symmetries are broken on the charged sectors of quantum electrodynamics [18], although the term "broken" is used here in a slightly different (weaker) sense.

An important feature of the breaking of continuous, internal symmetries and translation invariance is that it is always accompanied by the appearence of zero mass excitations (the Goldstone bosons, the phonons, respectively) which manifest themselves in the slow decay of correlations of suitably chosen observables; [6,20]. (The breaking of rotation invariance also implies the existence of correlations with slow decay but not the existence of Goldstone excitations).

Starting from our definition in Sect. 1, one can easily convince oneself - at least heuristically - that, in one and two dimensions (space dimensions, in equilibrium statistical mechanics; space-time dimensions in quantum field theory), continuous internal symmetries or translation invariance cannot be broken spontaneously or dynamically [6,20,21,22] (I recall the droplet argument, made rigorous in [22]), except in systems with interactions of extremely long range [23,24].

It may not be generally appreciated that we do not know any example of a system with translation invariant dynamics for which we can prove mathematically that the breaking of translation invariance occurs, (except in the rather unphysical, one-dimensional jellium model which exhibits a Wigner lattice or in idealized models of systems at positive density, but at zero temperature). Moreover, there are no known continuum models of liquid crystals for which we can rigorously establish directional long range order. These are serious gaps in the mathematical foundations of condensed matter physics.

Up until 1976 there were no known examples of models with a continuous, internal symmetry group for which spontaneous symmetry breaking, and hence the existence of Goldstone excitations, was established rigorously (except for the somewhat trivial spherical model of Berlin and Kac). That situation changed with the appearence of

[7], where a class of three - or higher dimensional lattice models of statistical mechanics, including the classical Heisenberg model, and the well known $\lambda|\vec{\varphi}|^4$-quantum field model in three space-time dimensions were discussed which exhibit spontaneous breaking of internal O(N) symmetries at suitably chosen values of the thermodynamic parameters, the coupling constant λ and the bare mass, respectively. The method in [7] involves a rigorous version of spin wave theory. It was extended in [25] to quantum-mechanical lattice models, including the quantum XY model and the Heisenberg anti-ferromagnet. The method was generalized considerably in [24]. However, the Heisenberg ferromagnet, for example, has so far resisted all attempts at a mathematically rigorous understanding. All these developments have been reviewed pedagogically e.g. in [3,8,9,26] , and I do not repeat that here.

A new method in the theory of phase transitions and continuous symmetry breaking has been introduced recently by Spencer and myself [27]. Unfortunately, it can only be applied to systems with an abelian symmetry group, so far, (although it is conceivable that one might recover the results on O(N) lattice σ-models of [7]). The method has the advantage of extending to abelian lattice gauge theories, permitting us to give fairly simple, new proofs of the results in [28,29]. It relies on a representation of abelian lattice spin systems or - field theories as gases of defects of co-dimension 2 carrying integer flux numbers. In three dimensions those defects are closely related to the Abrikosov vortices in a super conductor. They interact through long range forces. The broken symmetry appears when the defects have low effective activity , z , and form a dilute gas. The symmetry is restored when the defects condense. A heuristic discussion of the transition is based on an energy-entropy argument : The entropy , S, of a defect line (or-network) of length L is bounded by

$$S \leq c \cdot L \ ,$$

where c is a geometric constant, the energy is bounded by

$$E \geq c'|\varphi| \cdot L \ ,$$

where φ is the flux carried by the defect, i.e. $|\varphi| \geq 1$, and $c' > 0$. The effective activity z is therefore bounded above by

$$z \leq e^{-(\beta c' - c)L} \ , \ \beta = 1/kT \ ,$$

and is exponentially small in L for large $\beta(> c/c')$. Thus, for small temperatures T , the defects have small sizes and form a dilute gas. Hence the medium is ordered, and the symmetry is spontaneously broken. This argument is clearly reminiscent of the Peierls argument, e.g. [26], where one considers a gas of Bloch walls. It might have interesting applications to the theory of melting of solids. For a rigorous version see [27].

3. "Spontaneous breaking" of local, internal symmetries.

This topic is huge and less well understood than the breaking of global symmetries. It would require a series of lectures of its own. I therefore concentrate my attention on the discussion of a few specific aspects which are probably not typical for the whole circle of problems.

Systems admitting a local, internal symmetry group (i.e. general covariance under changes of internal "coordinates") are always described in the form of gauge (Yang-Mills) theories. Presently, the most widely used, non-perturbative ultraviolet regularization of gauge theories consists of putting these theories on a lattice. This preserves gauge-invariance, translation invariance and positivity of the metric in the physical Hilbert space, [30,31,32,33].

We now consider an example : The gauge group G is chosen to be $SU(2)$. The gauge field is denoted by $\underset{\sim}{g} = \{g_{xy}\}$, where xy is an arbitrary pair of nearest neighbors in $\mathbb{Z}^d$, and g_{xy} is an element of G formally given by

$$g_{xy} = P(e^{\int_x^y A_\mu(\xi)d\xi^\mu}), \quad \text{for all } xy.$$

We also introduce a Higgs field ,

$$\phi : x \in \mathbb{Z}^d \to \vec{\phi}_x \in \mathbb{E}^3,$$

with isospin 1 . Let χ be the spin 1/2 character of $SU(2)$, and U the spin 1 representation. The Euclidean functional measure which determines the vacuum state is given by

$$d\mu(g,\phi) \equiv Z^{-1} \prod_{(xy)} e^{\zeta(\vec{\phi}_x \cdot U(g_{xy})\vec{\phi}_y)} \cdot \prod_p e^{\beta\chi(g_{\partial p})} \prod_{(xy)} dg_{xy} \prod_x d\lambda(\vec{\phi}_x) \ , \tag{1}$$

$$d\lambda(\vec{\phi}) \overset{e.g.}{=} \exp[-\frac{\lambda}{4}|\vec{\phi}|^4 + \frac{\mu^2}{2}|\vec{\phi}|^2]d^3\phi \quad ,$$

$g_{\partial p} = \prod_{xy \subset \partial p}^{\circlearrowright} g_{xy}$, (p a unit square = plaquette), β, ζ and λ are positive constants. The r.s. of (1) is defined as the thermodynamic limit of the measures associated with finite sublattices, with arbitrary boundary conditions (b.c.) imposed at the boundary of each sublattice.

Let $h : x \to h_x$ be an arbitrary function from $\mathbb{Z}^d$ into G with the property that $h_x = 1$, except for finitely many sites x . We make the change of variables

$$g_{xy} \to h_x g_{xy} h_y^{-1} \equiv (g^h)_{xy} \ ,$$
$$\vec{\phi}_x \to U(h_x)\vec{\phi}_x \equiv (\vec{\phi}^h)_x \ . \qquad (2)$$

This change of variables leaves $d\mu(g,\phi)$ invariant, i.e. $d\mu(g,\phi) = d\mu(g^h,\phi^h)$, no matter what b.c. have been used to construct $d\mu$. Let

$$\langle\cdot\rangle \equiv \int(\cdot)d\mu(g,\phi) \ .$$

It clearly follows that gauge-dependent observables, like g_{xy} or $\vec{\phi}_x$, have zero expectation, i.e.

$$\langle g_{xy}\rangle = \langle\vec{\phi}_x\rangle = 0 \ , \qquad (3)$$

independent of the b.c. used in the construction of $\langle\cdot\rangle$. This is an immediate consequence of a definition of $\langle\cdot\rangle$ by means of the Dobrushin-Lanford-Ruelle equations, for example. It really expresses the triviality that if one does not fix a gauge by hand, the gauge is not fixed, and therefore gauge-dependent variables are averaged out when one computes their expectation (an observation made e.g. in [10,11]) .

Let us now fix a gauge and see whether we obtain a useful notion of "breaking of local symmetries". Gauge fixing is achieved by multiplying $d\mu$ by a function $F(g,\phi)$ with the property

$$\int F(g^h,\phi^h)\prod_x dh_x = 1 \ . \qquad (4)$$

Let $\langle\cdot\rangle_F$ be the expectation determined by $F(g,\phi)d\mu(g,\phi)$. For gauge-invariant observables, A ,

$$\langle A\rangle_F = \langle A\rangle \ . \qquad (5)$$

However, it is now possible, a priori, that

$$\langle\vec{\phi}_x\rangle_F \neq 0 \ ,$$

if suitable b.c. are imposed.

One possible, partial gauge fixing is to turn all Higgs variables, $\vec{\phi}_x$, parallel to the 3-axis, e_3 , of $\mathbb{E}^3$. (Choose F to be proportional to $\prod_x \delta(\vec{\phi}_x - |\vec{\phi}_x| e_3))$. Then

$$\langle\vec{\phi}_x\rangle_F = \langle|\vec{\phi}_x|\rangle_F \cdot e_3 \ , \ \langle|\vec{\phi}_x|\rangle_F > 0 \ , \qquad (6)$$

(no matter whether μ^2 is positive or negative). For this choice of F, $F(\phi)d\mu(g,\phi)$ has a residual, local invariance group : If in (2) every h_x leaves e_3 invariant $(U(h_x)e_3 = e_3, \forall x)$, hence belonging to a $U(1)$ subgroup of G, then

$$F(\phi^h)d\mu(g^h,\phi^h) = F(\phi)d\mu(g,\phi) . \tag{7}$$

Thus, in a sense the coupling of the gauge field to the Higgs field has broken the gauge group down to a residual $U(1)$, but since this happens no matter whether μ^2 is positive or negative, it does not provide a terribly useful notion of "spontaneous, local symmetry breaking". (This becomes particularly evident in a theory with a large gauge group G, e.g. $SU(3)$, and a Higgs field, ϕ, with the property that the action of G on the vector space, V, of possible values of $\vec{\phi}_x$ decomposes V into several inequivalent G-orbits). The above notion depends of course on our choice of F. It has been shown in [12] that, for some class of complete gauge fixings, F, including the temporal gauge, and arbitrary "symmetry breaking" b.c.

$$\langle\vec{\phi}_x\rangle_F = 0 . \tag{8}$$

(This follows either from a "spin-wave" argument related to the one in [22], or from the principle of "symmetry restoration via defects", such as instantons). In [12] Morchio, Strocchi and the author have therefore proposed a gauge-invariant description of the physics of a Higgs theory in the continuum limit, with the hope that this might lead to a useful notion of "spontaneous local symmetry breaking". In the example of the Georgi-Glashow model considered above, the appropriate, gauge invariant fields are

$$\vec{\phi}\cdot\vec{F}_{\mu\nu} \text{ (photon), and } \vec{\phi}\cdot\vec{\phi} \text{ (Higgs particle) .} \tag{9}$$

Moreover there is a gauge-covariant field

$$\pi_{\vec{\phi}}\vec{F}_{\mu\nu} \equiv N[|\vec{\phi}|^2\vec{F}_{\mu\nu}-\vec{\phi}(\vec{\phi}\cdot\vec{F}_{\mu\nu})] \quad (W^+ \text{ and } W^-) , \tag{10}$$

where N indicates normal ordering. From this field one may formally construct fields localized on curves (γ_{xy}) with given endpoints (x,y) :

$$(\pi_{\vec{\phi}}\vec{F}_{\mu\nu})(x)P[\exp\int_{\gamma_{xy}} A_\rho(\xi)\cdot d\xi^\rho](\pi_{\vec{\phi}}\vec{F}_{\sigma\lambda})(y) . \tag{11}$$

The conventional, gauge-dependent picture is recovered if $\vec{\phi}_x \propto e_3$. Note that the physical fields introduced in (9), (10) and (11) do not form any $SU(2)$ multiplets, and there is no reason why the masses of the photon and the W^+ and W^- boson ought to be degenerate. In fact, this theory is expected to have a non-perturbative phase in which there is only one massive, neutral vector particle (a massive photon), a

"Higgson" and neutral W^+-W^- bound states. In that phase the electric charge would be confined, (region I of Fig.1). At large renormalized values of ζ , μ^2 and β , the theory should however have a QED phase with a massless photon (coupled to the vacuum by $\vec{\phi}\cdot\vec{F}_{\mu\nu}$) , a massive (unstable) "Higgson", massive W^+ and W^- vector bosons and massive magnetic monopoles, (region II, Fig.1). When one speaks of "spontaneous breaking of SU(2)" in this theory one is thinking of phase II. The picture developped here can be tested in the lattice Georgi-Glashow model for which one expects the following phase diagram, $(\lambda,\mu^2 > 0 ; \lambda,\mu^2 \to \infty$, as $\zeta \to \infty)$:

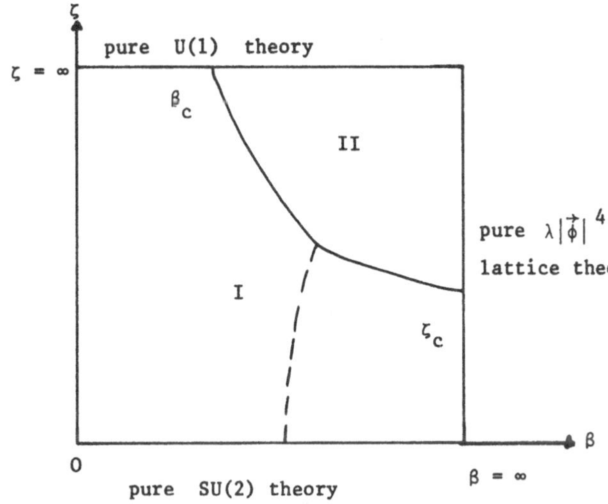

Fig. 1

I : confining phase/II : QED phase. The dashed line might correspond to a line of singularities of the electric string tension, [34] . The only rigorous results concern the existence and nature of the transitions on the lines $\beta = \infty$ and $\zeta = \infty$ (at ζ_c, β_c , resp.). See [7,28,27,33].

The above considerations extend to more general Higgs theories with the property that the stability group $H_{\vec{\phi}_x}$ of $\vec{\phi}_x \neq 0$ is conjugate to one subgroup H of the gauge group G , for all $\vec{\phi}_x$. The gauge fields corresponding to generators of the Lie algebra of $H_{\vec{\phi}_x}$, for all x , are the "electromagnetic and gluon fields", the remaining generators of the Lie algebra $\mathfrak{g}$ of G are the "broken generators" and should correspond to massive (H-neutral) bound states or massive vector bosons.

Things become problematic when there are several, inequivalent Higgs orbits, i.e. the abstract group corresponding to $H_{\vec{\phi}_x}$ depends non-trivially on $\vec{\phi}_x, \vec{\phi}_x \neq 0$. (Example : $G = SU(3)$, $\vec{\phi}$ in the adjoint representation, ...). Let $V_{eff}(\vec{\phi})$ be the effective Higgs potential, (including radiative corrections). Let $(\vec{\phi}_0)$ denote the orbit on which $V_{eff.}(\vec{\phi})$ takes its minimum, H_0 the corresponding (abstract) stability group, and $\mathfrak{h}_0$ the Lie algebra of H_0 . Let $\vec{\phi}_x$ be the Higgs field averaged over a ball centered at x of radius $\approx M^{-1}$, where M is a typical mass scale (a fluctuation length scale of ϕ) of the theory. Then with high probability

$$\overline{\vec{\phi}_x} \in (\vec{\phi}_o) . \tag{12}$$

Perturbation theorists then say that G is broken down to H_o, that the gauge fields corresponding to generators of $\mathfrak{h}_o$ (in the sense indicated above and in [12]) remain massless, while the ones corresponding to $\mathfrak{g} \ominus \mathfrak{h}_o$ acquire masses.

On a non-perturbative level, this prediction is probably wrong, as argued by Morchio and Strocchi [35] on the basis of ideas and results in statistical mechanics [36,37,38] and of [12] : Suppose $V_{eff.}(\vec{\phi})$ has a local minimum on an orbit $(\vec{\phi}_1)$, $\vec{\phi}_1 = \vec{\phi}_o + \delta\vec{\phi}_1$, with stability group $H_1 \subset H_o$. (In general $H_1 \not\subset H_o$, but we make this assumption for clarity). Let M_o^2, M_1^2 be the curvatures of $V_{eff.}$ transversal to $(\vec{\phi}_o)$, $(\vec{\phi}_1)$ at $\vec{\phi}_o, \vec{\phi}_1$, respectively. If

$$\Delta\alpha \equiv V_{eff.}(\vec{\phi}_1) - V_{eff.}(\vec{\phi}_o) + a(M_1^2 - M_o^2)M^2 >> 0 \quad , \tag{13}$$

for some positive constant $a \propto (\zeta^{-1})_{ren.}$, then $\overline{\vec{\phi}_x}$ is close to $(\vec{\phi}_o)$ predominantly. (If (13) fails it may happen that $\overline{\vec{\phi}_x}$ is close to $(\vec{\phi}_1)$, predominantly, even if $V_{eff.}(\vec{\phi}_1) > V_{eff.}(\vec{\phi}_o)$) . However, with some probability (vanishing in perturbation theory, but positive non-perturbatively) there appear "bubbles, B , of the false vacuum" such that $\overline{\vec{\phi}_x}$ is close to $(\vec{\phi}_1)$, for $x \in B$. The effect of these bubbles on the physics of such a theory can be estimated by a Peierls (action-entropy) argument [36] and a study of mass generation [38]. (I follow a presentation in [37]): First we must estimate the probability p of the event E_x that $\overline{\vec{\phi}_x}$ is close to $(\vec{\phi}_1)$, i.e. that x (e.g. the origin) belongs to a bubble $B \supseteq \{y \,|\, |y-x| \leq M^{-1}\}$ of false vacuum. We choose b.c. such that $\overline{\vec{\phi}_z} \in (\vec{\phi}_o)$, as $|z| \to \infty$. A connected piece, Γ, of the boundary of a bubble is called a contour (or phase boundary). For the event E_x to occur it is necessary that there be a contour Γ separating $\{y \,|\, |y-x| \leq M^{-1}\}$ from ∞, as follows from our definitions of $\overline{\vec{\phi}_x}$ and of contours. The action of a bubble B such that $\partial B \supset \Gamma$ is bounded below by $A(|\Gamma|)$, where

$$A(|\Gamma|) \geq \sigma M^{-3}|\Gamma| + \Delta\alpha \cdot M^{-1}|\Gamma| \quad . \tag{14}$$

Here σ is a constant $\propto (\zeta\mu^2\lambda^{-1})_{ren.}$ and $|\Gamma|$ is the volume of Γ . The first term is a surface term, the second term a lower bound for a volume term. The precise dependence of M , a and σ on coupling constants is not known, presently. If

$$A(|\Gamma|)|\Gamma|^{-1} >> 1 \tag{15}$$

the statistical weight of a contour Γ is bounded above by $\exp[-A(|\Gamma|)]$. Therefore the probability p for E_x to occur is bounded by

$$p \leqq \sum_{\Gamma}^{x} \exp[-A(|\Gamma|)] \quad ^{1)} \quad , \tag{16}$$

where $\sum_{\Gamma}^{x}$ ranges over all contours Γ of volume $|\Gamma| = \text{const.}M^{-3}n$, $n = 1,2,3,\ldots$, surrounding $\{y \,|\, |y-x| \leqq M^{-1}\}$. The number of such contours with given volume, $|\Gamma| = \text{const.}M^{-3}n$, is bounded by

$$e^{cn}, \quad (c \approx O(1) \text{ is a geometrical constant}) . \tag{17}$$

From (14), (16) and (17) we conclude that

$$p \ll 1 \quad \text{if (15) holds.} \tag{18}$$

By the results of [38,35] one then expects that gauge fields corresponding to generators in $\mathfrak{g} \ominus \mathfrak{h}_0$ (in the sense of [12]) acquire masses $\propto |\vec{\phi}_0|$, while gauge fields corresponding to the generators of $\mathfrak{h}_0 \ominus \mathfrak{h}_1$ acquire masses

$$\propto \sqrt{p}\,|\delta\vec{\phi}_1| . \tag{19}$$

Similar considerations apply to Fermion masses. Thus - if there are no further local minima giving rise to other bubbles - G is really "broken down" to H_1, rather than to the larger group H_0. Moreover, there is no elementary Higgs field causing the "breaking" from H_0 down to H_1, [35]. Finally, one does not expect any dynamical monopoles with charges labelled by $\pi_2(H_0/H_1)$, (π_k = k^{th} homotopy group), but only ones with charges labelled by $\pi_2(G/H_0) \cong \pi_1(H_0)$, as argued in [37].

Finally, we should mention that the applicability of conventional perturbation theory, based on the (generally incorrect) assumption that $\langle\vec{\phi}_x\rangle_F \neq 0$, to Higgs gauge theories has been discussed in [12], with the result that the deviations can generally be expected to be entirely non-perturbative.

This ends our discussion of the notion of "spontaneous breaking of local (gauge) symmetries" : In abstracto, it is somewhat vague and misleading. It must be understood dynamically, and one should be aware of the fact that non-perturbative effects generally alter the conventional interpretation. Such effects, together with the requirements of renormalizability, some form of asymptotic freedom and the requirement that there exist an "unbroken" $SU(3)_c \times U(1)$ may be useful guides for model builders.

---

1) A lower bound on p is more difficult to derive, see [37].

4. Renormalization group ideas.

We consider a class of physical systems which can be described by a family (algebra), $\mathfrak{A}$, of local "observables", e.g. Euclidean fields, in quantum field theory (QFT), or spin fields, in statistical mechanics (SM), and some space, X , of time-translation invariant states, e.g. Euclidean vacuum functionals in QFT , equilibrium states in SM . Let $A \in \mathfrak{A}$. By A_x we denote the translate of A by a vector x in space-imaginary time (QFT) , or space (SM) . Let $\rho \in X$ be a state characterizing a specific physical system. Question : How do correlations,

$$\rho(A_x \cdot B_y) \ , \ A,B \ \text{ in } \mathfrak{A} \ ,$$

behave, as $|x-y| \to \infty$, i.e. in the (infrared) scaling limit ? (In a continuum system one may also be interested in the behaviour of $\rho(A_x \cdot B_y)$, as $|x-y| \to 0$: the short distance, or ultraviolet limit. We focus our attention on the scaling limit). In order to answer that question, one tries to construct functions, $\alpha_A(\theta)$, depending on $A \in \mathfrak{A}$ and on a scale parameter θ , such that

$$G_{A,B}(x,y) \equiv \lim_{\theta\to\infty} \alpha_A(\theta)\alpha_B(\theta)\{\rho(A_{\theta x} \cdot B_{\theta y}) - \rho(A_{\theta x})\rho(B_{\theta y})\} \qquad (20)$$

exists. (My discussion is slightly oversimplified at this point, since one often chooses ρ on the r.s. of (20) to depend on θ , as well, such that ρ_θ approaches a critical state, as $\theta \to \infty$). In order to find $\alpha_A(\theta)$, $A \in \mathfrak{A}$, and other quantities of interest at large distances, one tries to determine the large scale effective dynamics, by integrating out fluctuations on a sequence of increasing length scales. One popular scheme to accomplish that is the Kadanoff "block spin transformations". Abstractly, they can be described as a non-linear transformation, τ , acting on $X \times \mathfrak{A}$:

$$\tau : (\rho,A) \to \tau(\rho,A) \equiv (\rho_\tau, A_\tau) \ , \qquad (21)$$

$$\text{with } \rho_\tau(A_\tau) = \rho(A) \ ^{2)} \ ,$$

such that each application of τ increases the scale of effective fluctuations, i.e. transforms a dynamics on a given scale into an effective dynamics on the next larger scale. In order to answer the question raised at the beginning by means of such a scheme one must study the manifold M_m of fixed points of τ :

$$\rho^* \in M_m \subset X \ \text{ iff } \ \rho^*_\tau = \rho^* \ . \qquad (22)$$

---

2) or $\tau = \tau_\theta$, with $\rho(A_{\theta x} \cdot B_{\theta y}) = \text{const.}\rho_{\tau_\theta}(A_x \cdot B_y)$. (21')

Under suitable hypotheses on the properties of τ, one can decompose X in the vicinity of some $\rho^* \in M_m$ into a stable manifold, $M_s(\rho^*)$, and an unstable manifold, $M_u(\rho^*)$:

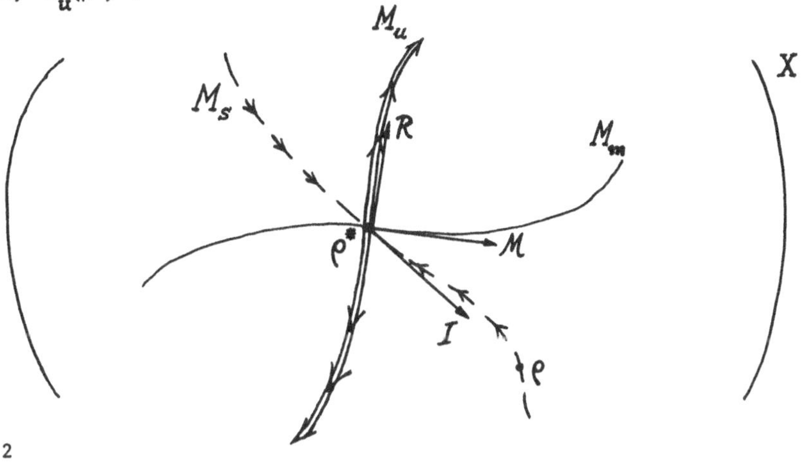

Fig. 2

States on $M_s(\rho^*)$ are driven towards ρ^*, states on $M_u(\rho^*)$ are driven away from ρ^*, under the action of τ. The tangent space, R, to $M_u(\rho^*)$ at ρ^* is the linear space spanned by eigenvectors of $D\tau_{\rho^*}$, the linearization of τ at ρ^*, corresponding to eigenvalues of modulus > 1. R is called the space of "relevant perturbations". The space, I, of "irrelevant perturbations" is defined by replacing τ by τ^{-1} in the definition of R, and the space, M, of "marginal perturbations" is the tangent space to M_m at ρ^*. Let $\rho \in M_s(\rho^*)$. One argues that the functions $\alpha_A(\theta)$ are computable in terms of A and of the rate of approach of $\underbrace{(\cdots(\rho_\tau)_\tau)_\tau\cdots}_{n \text{ times}}$ to ρ^*. (See (21), (21')).

The point of interest to us is now the following : It may happen that the fixed point ρ^* has a larger symmetry group than a state ρ on $M_s(\rho^*)$. This entails that the scaled correlations, $G_{A,B}(x,y)$, exhibit a larger symmetry than the original correlations $\rho(A_x \cdot B_y)$. If this happens we speak of asymptotic enhancement of symmetry, or of the (dynamical) generation of asymptotic symmetries. It is quite irrelevant in this general discussion, whether the symmetry in question is internal or spatial, global or local (i.e. gauged).

One might argue that the concept of symmetry enhancement is only interesting for physics if it has some stability properties. Let G be some (global or local, internal or spatial) symmetry group, and let H be a subgroup of G. Consider a G-invariant fixed point, ρ^*, of τ, (τ is assumed to have suitable smoothness properties). Suppose that the H-invariant subspace of M coincides with the G-invariant subspace of M. Then, in some vicinity, N, of ρ^*, every H-invariant fixed point of τ is also G-invariant. Thus, all states in $\bigcup_{\tilde{\rho}/N} M_s(\tilde{\rho})$, where $\bigcup_{\tilde{\rho}/N}$ ranges over all H-invariant fixed points, $\tilde{\rho}$, of τ in N, are driven towards G-invariant fixed points. Moreover if the H-invariant subspace of $M \oplus R$ coincides with the G-invariant subspace of $M \oplus R$, at ρ^*, then for some neighborhood N of ρ^*, the H-invariant subspace of marginal and relevant perturbations of a H-invariant fixed point $\tilde{\rho} \in N$ is also G-invariant, (i.e. H-invariant states near ρ^* tend to approach G-invariant states under the action of τ). This is the desired stability of our concept. The concept of symmetry enhancement has been described e.g. in [14], (see also Dürr's notes). The first rigorous study of models exhibiting this phenomenon (e.g. the $\mathbb{Z}_N$-models, see Sect. 5) probably appeared in [27,39]. An abstract discussion very similar to the one presented here appeared subsequently in [16] (which inspired the present section).

5. Symmetry enhancement : Generation of asymptotic, global and local symmetries.

In this final section we sketch very briefly some examples of the phenomena described at the end of Sect. 4 and point out why "symmetry enhancement" might be the right concept permitting us to decide whether a local (gauge) symmetry in some gauge theory is "spontaneously broken", or not.

First, we consider the $\mathbb{Z}_N$ spin models on the square lattice, $\mathbb{Z}^2$, $N = 5,6,7, \ldots$. The classical spin, $\vec{S}_x$, at a site $x \in \mathbb{Z}^2$ is given by

$$\vec{S}_x = (\cos\theta_x, \sin\theta_x) , \quad \theta_x = \frac{2\pi n_x}{N} , \quad n_x = 0,\ldots,N-1, \ \forall x .$$

The equilibrium state of the model at inverse temperature β is given by the measure

$$d\mu_\beta^{(N)}(\theta) = [Z_\beta^{(N)}]^{-1} \exp[\beta \sum_{xy} \cos(\theta_x - \theta_y)] , \tag{23}$$

where xy is an arbitrary pair of nearest neighbors. This measure is the limit of the measures

$$d\mu_{\beta,h}(\theta) = Z_{\beta,h}^{-1} \exp[\beta \sum_{xy} \cos(\theta_x - \theta_y)] \prod_x e^{h\cos(N\theta_x)} d\theta_x , \tag{24}$$

as $h \to \infty$; ($d\theta$ = Lebesgue measure on unit circle). The classical XY- , or rotator model corresponds to $h = 0$. For $h > 0$, the measure $d\mu_{\beta,h}$ and $d\mu_\beta^{(N)}$ have a discrete, global symmetry group (generated by $\mathbb{Z}_N$ and reflections) while the rotator ($h = 0$) has a continuous, global symmetry group. In [39] Spencer and the author have shown that, for all $h \in [0,\infty]$ and $N \geq N_o$, where N_o is a suffiently large integer independent of h , there exists an interval $[\underline{\beta}(h,N) , \overline{\beta}(h,N)]$ of values of β which are all critical points and at which the correlation length of the spin systems described by $d\mu_{\beta,h}$ is infinite. Moreover,

$$\underline{\beta}(h,N) \leq \underline{\beta}(0,N) \equiv \beta_c(\text{rotator}) < \infty ,$$

$$\overline{\beta}(h,N) \to \infty , \text{ as } h \to 0 \text{ or } N \to \infty .$$

We have constructed an infinite sequence of renormalization transformations which drive $d\mu_{\beta,h}$ towards a U(1)-invariant state $d\mu^*_{\beta,h}$, for all $h \in (0,\infty]$, $N \geq N_o$, and all $\beta \in (\beta_1,\beta_2)$, with $\underline{\beta}(h,N) \leq \beta_1 < \beta_2 \leq \overline{\beta}(h,N)$. Thus, asymptotically, the discrete symmetry of the $\mathbb{Z}_N$ models is enhanced to a continuous symmetry. We conjecture that for each $h \in (0,\infty]$ and each $\beta \in (\underline{\beta}(h,N),\overline{\beta}(h,N))$, $N \geq N_o$, there exists some $\beta' \equiv \beta'(\beta,h) \geq \beta_c$ (rotator) such that spin correlations in $d\mu_{\beta,h}$ and in $d\mu_{\beta'}$ have identical (long distance) scaling limits, (although this does not quite follow from our construction).

In [27] we have established similar results for the QED phases [29] of the $\mathbb{Z}_N$ lattice gauge theory in four dimensions : Local $\mathbb{Z}_N$-invariance is asymptotically enhanced to local U(1)-invariance.

Recently, we have also examined examples of non-abelian gauge theories coupled to some Higgs fields (not transforming under the fundamental representation) for which we argue that, for suitable choices of the coupling constants $\beta,\zeta,\lambda,\mu^2$, the theory is in the same (long distance) "universality class" as the corresponding pure Yang-Mills theory (for some $\beta' = \beta'(\beta,\zeta,\ldots)$, $\zeta = 0$, $\phi = 0$), if only gauge field expectations are considered. In such a case one could say that the matter fields leave the full gauge group "unbroken". (In the opposite case it would be appropriate to speak of "local symmetry breaking"). It would be interesting to study symmetry enhancement at short distances in continuum grand unified theories.

More standard examples of symmetry enhancement which are, however, not very well understood mathematically are :

- Restoration of full Euclidean invariance of correlations of lattice theories in the scaling limit (as $\beta \nearrow \beta_c$, where β_c is a critical point).

- Restoration of translation invariance above the roughening temperature in the three-dimensional Ising model or in a lattice gauge theory, [34].

Problems of symmetry enhancement are typically very involved, technically, so that we cannot present any details here.

References.

1. H. Weyl, "Symmetry", Princeton, N.J. : Princeton University Press, 1952.

2. S. Coleman, Secret Symmetry : An Introduction to Spontaneous Symmetry Breakdown and Gauge Fields, Erice Lectures 1973, A. Zichichi (ed.).

3. J. Fröhlich, Bull. Amer. Math. Soc. 84, 165, (1978).

4. L. Michel, Reviews of Modern Physics 52, 617, (1980).

5. J. Goldstone, Nuovo Cimento 19, 15, (1961);
Y. Nambu and G. Jona-Lasinio, Phys. Rev. 122, 345, (1961), 124, 246, (1961).

6. H. Ezawa and J.A. Swieca, Commun. math. Phys. 5, 330, (1967).

7. J. Fröhlich, B. Simon and T. Spencer, Commun. math. Phys. 50, 79, (1976).

8. J. Fröhlich and T. Spencer, in "New Developments in Quantum Field Theory and Statistical Mechanics", M. Lévy and P. Mitter (eds.), New York & London : Plenum, 1977.

9. J. Fröhlich, Acta Physica Austrica Suppl. XV, 133, (1976).

10. S. Elitzur, Phys. Rev. D12, 3978, (1975).

11. G.F. De Angelis, D. De Falco and F. Guerra, Phys. Rev. D17, 1624, (1978).

12. J. Fröhlich, G. Morchio and F. Strocchi, Phys. Letts. 97B, 249, (1980); Nucl. Phys. B, in press.

13. K. Wilson and J. Kogut, Physics Reports 12C, N°2, 76, (1974).
K. Wilson, Rev. Mod. Phys. 47, N°4
L. Kadanoff, A. Haughton and M. Yalabik, J. Stat. Phys. 14, N°2, 171, (1976).
G. Jona-Lasinio, Nuovo Cimento 26B,99, (1975).
S. Ma, Rev. Mod. Phys. 46, N°4, 589, (1973).
P. Bleher and Ja.Sinai, Commun. math. Phys. 33, 23, (1973), 45, 247, (1975).
G. Jona-Lasinio, in "New Developments..." (see ref. 8).
M.E. Fisher, Rev. Mod. Phys. 46, N°4, 597, (1974).

14. D. Foerster, H.B. Nielsen, M. Ninomiya, Phys. Lett. 94 B, 135, (1980).
J. Iliopoulos, D.V. Nanopoulos and T.N. Tomaras, Phys. Lett. 94 B, 141, (1980).
Réf. 39 (Sect. 7); ref. 27; ref. 16.
K. Cahill and P. Denes, Preprint, Univ. New Mexico : UNMTP-81/020.

15. Refs. 39 and 27;
J. Fröhlich and T. Spencer, Phase Diagrams and Critical Properties of Classical Coulomb Systems, Erice 1980.

16. C. Newman and L. Schulman,"Asymptotic Symmetry : Enhancement and Stability," submitted to Phys. Rev. Letters.

17. L. Michel and L. Radicati, Ann. Phys. (NY) 66, 758, (1971).
D. Kastler et al., Commun. math. Phys. 27, 195, (1972).

18. J. Fröhlich, G. Morchio and F. Strocchi, Ann. Phys. (N.Y.) 119, 241, (1979), Phys. Lett. 89 B, 61, (1979).

19. Ph. Martin, Preprint, EPF-Lausanne, 1981.

20. N.D. Mermin, J. Math. Phys. 8, 1061, (1967).
J. Phys. Soc. Japan, Suppl. 26, 203, (1969).

21. S. Coleman, Commun. math. Phys. 31, 259, (1974). See also ref. 6.

22. J. Fröhlich and C. Pfister, Commun. math. Phys. (1981).

23. H. Kunz and C. Pfister, Commun. math. Phys. 46, 245, (1976).

24. J. Fröhlich, R. Israel, E.H. Lieb and B. Simon, Commun. math. Phys. 62, 1, (1978).

25. F. Dyson, E.H. Lieb and B. Simon, J. Stat. Phys. 18, 335, (1978).

26. E.H. Lieb, in "Mathematical Problems in Theoretical Physics", G.F. Dell'Antonio, S. Doplicher and G. Jona-Lasinio (eds.), Springer Lecture Notes in Physics, Berlin-Heidelberg-New York : Springer Verlag, 1978.

27. J. Fröhlich and T. Spencer, "Massless Phases and Symmetry Restoration....", Commun. math. Phys., to appear.

28. A. Guth, Phys. Rev. D21, 2291, (1980).

29. S. Elitzur, R. Pearson and J. Shigemitsu, Phys. Rev. D19, 3698, (1979).

30. K. Wilson, Phys. Rev. D10, 2445, (1974).

31. K. Osterwalder and E. Seiler, Ann. Phys. (NY) 110, 440, (1978).

32. D. Brydges, J. Fröhlich and E. Seiler, Ann. Phys. (NY) 121, 227, (1979).

33. E. Seiler, "Gauge Theories as a Problem of Constructive Quantum Field Theory and Statistical Mechanics", Springer Lecture Notes in Physics, to appear.

34. H. van Beijeren, Commun. math. Phys. 40, 1, (1975), Phys. Rev. Lett. 38, 993, (1977);
ref. 39, (Sect. 7); C. Itzykson, M.E. Peskin and J.-B. Zuber, Phys. Lett. 95 B, 259, (1980);
A. Hasenfratz, E. Hasenfratz and P. Hasenfratz, Nucl. Phys. B180, 353, (1981);
M. Lüscher, DESY Preprint 1980.

35. G. Morchio and F. Strocchi, Phys. Lett. 104 B, 277, (1981).

36. J. Glimm, A. Jaffe and T. Spencer, Commun. math. Phys. 45, 203 (1975);
R. Dobrushin and S. Schlosman, Preprint 1981.

37. J. Fröhlich, "The Statistical Mechanics of Defect Gases", unpublished.

38. D. Brydges and P. Federbush, Commun. math. Phys. 62, 79, (1978);
D. Brydges, J. Fröhlich and T. Spencer, "The Random Walk Representation of Classical Spin Systems and Correlation Inequalities", Commun. math. Phys., to appear.

39. J. Fröhlich and T. Spencer, "The Kosterlitz-Thouless Transition in Two-Dimensional Abelian Spin Systems and the Coulomb Gas", Commun. math. Phys. to appear.

SPONTANEOUS BREAKING OF SUPERSYMMETRY

Bruno Zumino
Lawrence Berkeley Laboratory
and
Department of Physics
University of California
Berkeley, California 94720 U.S.A.

1. INTRODUCTION

There has been recently a revival of interest in supersymmetric gauge theories, stimulated by the hope that supersymmetry might help in clarifying some of the questions which remain unanswered in the so called Grand Unified Theories and in particular the gauge hierarchy problem. In a Grand Unified Theory[1] one has two widely different mass scales: the unification mass $M \simeq 10^{15}$GeV at which the unification group (e.g. SU(5)) breaks down to SU(3) × SU(2) × U(1) and the mass $\mu \simeq 100$ GeV at which SU(2) × U(1) is broken down to the U(1) of electromagnetism. There is at present no theoretical understanding of the extreme smallness of the ratio μ/M of these two numbers. This is the gauge hierarchy problem.

There is a more technical aspect to the hieracrchy problem.[2] In a Grand Unified Theory the two mass scales come from the vacuum expectation values of two Higgs fields, which in turn are related to the parameters entering the Higgs potential. For the gauge hierarchy to emerge, some Higgs fields must have a small mass close to μ while others must have a large mass close to M. This requires a "fine tuning" of the parameters of the Higgs potential which, however, is in general unstable under radiative corrections. As recently emphasized by Witten,[3] there are special properties of supersymmetric theories which could help in this connection, namely the absence of renormalization of some of the parameters entering the Lagrangian, for instance masses and scalar couplings.[4-7] More simply, one could hope that, in a supersymmetric theory, the smallness of a scalar mass is guaranteed by the smallness of the mass of its spinor superpartner, which in turn is guaranteed by an approximate chiral invariance. Of course, a solution of the numerical hierarchy puzzle itself will require more than these special naturalness properties of supersymmetric theories (called sometimes in jest "supernaturalness") and can be found perhaps in non-perturbative breaking of supersymmetry.[3]

I shall not review here the numerous recent papers attempting to construct realistic models of supersymmetric gauge theories. As in previous work mostly by Fayet, these papers use N = 1 supersymmetry and do not attempt unification with gravity. Supersymmetry must of course be broken, the scale of supersymmetry breaking being at least 15 to 20 GeV for consistency with experiment. In this lecture I shall attempt to review the various mechanisms for spontaneous supersymmetry breaking[9] in gauge theories. Most of the discussion will be concerned with the tree approximation but what is presently known about radiative correction will also be reviewed.

2. SCALAR-SPINOR SUPERMULTIPLETS

The supersymmetric Lagrangian[10] for n interacting chiral (spin 0 - spin $\frac{1}{2}$) supermultiplets ϕ_i $(i = 1,2,\ldots n)$ is the sum of the kinetic term plus an interaction which can be derived from a single function of the ϕ_i, which we shall call the superpotential. For a renormalizable theory the superpotential is a cubic polynomial

$$f(\phi) = a + b_i\phi_i + \frac{1}{2} m_{ij}\phi_i\phi_j + \frac{1}{3!} g_{ijk}\phi_i\phi_j\phi_k \tag{2.1}$$

(sum over repeated indices). The chiral superfields ϕ_i are complex and so are their scalar components A_i and the corresponding auxiliary fields F_i. The part of the Lagrangian which describes the scalar interactions is

$$\mathcal{L}_{S.I.} = F_i\bar{F}_i + F_i \frac{\partial f}{\partial A_i} + \bar{F}_i \overline{\frac{\partial f}{\partial A_i}} . \tag{2.2}$$

The equations of motion obtained by varying (2.2) with respect to F_i and $\bar{F}_i$ are

$$\bar{F}_i = - \frac{\partial f}{\partial A_i} \tag{2.3}$$

and their complex conjugates. Substituting (2.3) into (2.2) and changing the sign, one obtains the tree approximation scalar potential

$$V = \frac{\partial f}{\partial A_i} \overline{\frac{\partial f}{\partial A_i}} \geqslant 0. \tag{2.4}$$

The scalar potential is non-negative. If it is equal to zero at its minimum, supersymmetry is exact, if it is positive at its minimum supersymmetry is spontaneously broken. If supersymmetry is exact, the equations

$$\frac{\partial f}{\partial A_i} = 0 \tag{2.5}$$

must have a common solution $A_i = \overset{o}{A}_i$. Since (2.5) are n quadratic equations in n complex unknowns, in general they will have 2^n solutions but in special cases they may have no solutions[10,11] or they may have a continuous infinity of solutions, in which case there are massless scalars in the theory. It is not difficult to construct examples for all three situations. When there are more than one solution, any of them is equally acceptable as a vacuum. It can be shown[4,5] that, in perturbation theory, higher order corrections do not renormalize the second and third term in the right hand side of (2.2): the superpotential is unmodified by higher order corrections. Furthermore[12-18] higher order corrections cannot induce spontaneous breaking of supersymmetry nor can they remove the degeneracy when there are several acceptable zero-energy vacua at the tree approximation. If there are massless scalars they remain massless.

The reason for all this is that, for x-independent fields, all higher order corrections to the scalar interaction (2.2) have the form[12, 7]

$$\mathcal{L}'_{S.I.} = F_i\bar{F}_j h_{ij}(F, \bar{F}, A, \bar{A}), \tag{2.6}$$

where h_{ij} is a hermitean matrix, function of the indicated variables, which can be calculated in perturbation theory. One sees immediately that only the first term in the right hand side of (2.2) is renormalized and the wave function renormalization matrix is

$$\delta_{ij} + h_{ij}(0, 0, 0, 0). \tag{2.7}$$

If we add (2.6) to (2.2), the equations of motion for F_i and $\bar{F}_i$ become

$$\bar{F}_i + \frac{\partial f}{\partial A_i} + \bar{F}_j h_{ij} + F_j\bar{F}_k \frac{\partial h_{jk}}{\partial F_i} = 0, \tag{2.8}$$

plus the complex conjugates. On the other hand, the equations for A_i become

$$F_j \frac{\partial^2 f}{\partial A_i \partial A_j} + F_j\bar{F}_k \frac{\partial h_{jk}}{\partial A_i} = 0. \tag{2.9}$$

Clearly, a solution $A_i = \overset{o}{A}_i$ of (2.5), together with $F_i = 0$, satisfies both (2.8) and (2.9). The sum of (2.2) and (2.6) vanishes for those values of A_i and F_i. Therefore, a possible vacuum at the tree approximation is a possible vacuum to all orders.[18,19] Observe that, since the energy cannot become negative (this is a consequence of the supersymmetry algebra) all the solutions of (2.5) give true minima to all orders.

Let us now consider spontaneous breaking of supersymmetry.[20] At the tree approximation this means that (2.5) have no solutions and the F_i cannot all vanish. In this case one can show that the potential (2.4) cannot be "field-confining". We define a potential to be field-confining when it tends to infinity if A_i tends to infinity so that the fields A_i cannot become arbitrarily large. More precisely

$$V \to \infty \tag{2.10}$$

if

$$|A|^2 \equiv A_i\bar{A}_i \to \infty . \tag{2.11}$$

For a non confining potential, let us assume that there exists a positive number p such that

$$V \geqslant p > 0. \tag{2.12}$$

This excludes unphysical potentials which tends to zero when one of the scalar fields tends to infinity. Then one can show that the determinant of the second derivatives of the superpotential vanishes indentically[21] in A_i

$$\det \frac{\partial^2 f}{\partial A_i \partial A_j} \equiv 0, \tag{2.13}$$

and the matrix has therefore at least one vanishing eigenvalue. Calculated at the minimum of the potential (2.4), this matrix is the spinor mass matrix, which must have a zero eigenvalue corresponding to the Goldstone spinor of spontaneously broken supersymmetry (see (2.17) below). The fact that it has a zero eigenvalue for all values of the scalar fields A_i implies special properties of the Yukawa couplings.

It can be shown that the components of the eigenvector corresponding to zero eigenvalue

$$\frac{\partial^2 f}{\partial A_i \partial A_j} v_j(A) = 0 \tag{2.14}$$

are polynomials in A_i (independent of $\bar{A}_i$). The corresponding differential operator applied to the potential (2.4) gives zero identically

$$v_j(A) \frac{\partial V}{\partial A_j} = 0 \tag{2.15}$$

and the same is true of the complex conjugate differential operator. Along the complex curves defined by the differential equations

$$\frac{dA_i}{dt} = v_i(A) \tag{2.16}$$

the potential is constant. Assume that the potential reaches its minimum for a finite value of the scalar fields. Given any minumum of the potential, there is one of these curves (2.14) passing through it, which implies the presence of a complex massless scalar (actually these valleys of minima extend to infinity). Observe that, from (2.4),

$$\frac{\partial V}{\partial A_i} = \frac{\partial^2 f}{\partial A_i \partial A_j} \overline{\frac{\partial f}{\partial A_j}}. \tag{2.17}$$

At a minimum this must vanish (together with its complex conjugate). Therefore, if there is only one vanishing eigenvalue, one must have there the proportionality

$$v_i(A) \propto \overline{\frac{\partial f}{\partial A_i}} \tag{2.18}$$

between a polynomial vector whose components are functions of A_i only and one whose components are functions of $\bar{A}_i$ only. All these general results can be easily checked in the special examples of spontaneous breaking of supersymmetry discussed in Refs. 10, 11.

The necessity of massless scalars in addition to the Goldstone spinor may seem strange, but it is a property of the tree approximation only. When supersymmetry is spontaneously broken, the radiative corrections, which still have the form (2.6), change the situation in an essential way, because the F_i do not vanish. Already at the one-loop level the degeneracy of the valley of minima is lifted[22,23] and in general one has only one absolute minimum and no massless scalars. The potential increases with the scalar fields so that the minimum is for relatively small values of the fields. The value of the potential at the minimum also changes in the one loop approximation. All this has been verified in several special examples.[22]

3. SUPERSYMMETRIC GAUGE THEORIES

We consider now the case when there are gauge fields present. If the gauge group is simple, the tree approximation scalar Lagrangian (2.2) must be complemented by

$$\frac{1}{2}(D^a)^2 + gD^a\bar{A}T^aA \qquad (3.1)$$

where the scalar fields A now belong to some representations of the guage group, T^a are the matrices which represent the generators of the group and g is the gauge coupling constant. If the gauge group is semi-simple, one has the sum of a number of terms like (3.1). If the gauge group contains U(1) factors, each U(1) factor contributes to the sum a term of the form

$$\frac{1}{2}D^2 + g_1 D\,\bar{A}YA + \xi D, \qquad (3.2)$$

where g_1 is the U(1) gauge coupling constant and Y the U(1) charge of the scalar fields. The ξD term is the Fayet-Iliopoulos term,[24] which can induce spontaneous supersymmetry breaking. Eliminating the field D^a through its equations of motion, (3.1) gives a term of the form

$$-\frac{1}{2}g^2(\bar{A}T^aA)^2, \qquad (3.3)$$

while (3.2) gives rise to

$$-\frac{1}{2}(g_1\bar{A}YA + \xi)^2. \qquad (3.4)$$

In the scalar potential the negatives of (3.3) and (3.4) enter.

So, when gauge fields are present, the scalar potential consists of (2.4) plus a sum of terms like the negatives of (3.3) and (3.4)

$$V = \frac{\partial f}{\partial A_i}\overline{\frac{\partial f}{\partial A_i}} + \Sigma\frac{1}{2}g^2(\bar{A}T^aA)^2 + \Sigma\frac{1}{2}(g_1\bar{A}YA + \xi)^2 \geq 0. \qquad (3.5)$$

If for a value $A_i = \overset{o}{A}_i$ of the scalar fields the potential in (3.5) vanishes, supersymmetry is exact in the tree approximation. Again one can show[13-16,18] that higher order corrections will not break supersymmetry and will not remove any degeneracy which may exist in the tree approximation. Since the effective potential, to any order, is determined by a knowledge of the renormalization group functions[25] this fact can be related to the special properties of supersymmetric gauge theories with respect to renormalization. The only renormalization constants needed are:[6] a wave function renormalization for each chiral superfield, a wave function renormalization for each gauge superfield and a gauge coupling renormalization for each gauge coupling constant. No separate mass and scalar coupling renormalizations are necessary, which gives relations among the renormalization group functions. Of course, those superfields which belong to the same irreducible representation of the gauge group have the same wave function renormalization.

If the chiral superpotential gives rise to spontaneous breaking of supersymmetry in the tree approximation, which means that (2.5) have no solution, the presence of gauge fields does not change the fact that supersymmetry is spontaneously broken, since the additional terms in (3.5) are positive. On the other hand, let us assume that the first term in the right hand side of (3.5) vanishes for some value $A_i = \overset{o}{A}_i$ of the scalar fields. We distinguish several cases.

Let us first consider the case when there are no U(1) factors, so that the last term in the right hand side of (3.5) is missing. If $\overset{o}{A}_i = 0$, the second term vanishes: supersymmetry is exact. If not all the $\overset{o}{A}_i$ vanish, the second term in (3.5) does not vanish in general, however this does not necessarily mean that supersymmetry is broken. The superpotential f(A) is invariant under the semi-simple gauge group; as it was first pointed out by Ovrut and Wess,[26] this means that f(A) is also invariant under the complex extension of the group (same generators, but the parameters are allowed to be complex instead of being restricted to be real). This complex invariance can be used to find other values of A_i where the first term in (3.5) still vanishes. The second term is not invariant under the complex extension of the group and one can show that it can be transformed to zero by using a transformation of the complex extension of the group. In conclusion, for a semisimple gauge group, if the chiral part of the scarlar potential (the f dependent part) reaches the value zero for some value of the scalar fields, even if the gauge term does not vanish at that point, one can find another value of the scalar fields where both terms vanish. This is then a true minimum and supersymmetry is exact.[27]

This result is also valid if there is one U(1) factor even with non vanishing ξ, provided the chiral part of the potential vanishes for non zero scalar field. In this case supersymmetry cannot be spontaneously broken if it is not already broken by the chiral superpotential. However, if there is more than one U(1) factor, one cannot prove an analogous result in general, although, if there are enough chiral supermultiplets in the theory the statement tends to be correct anyway in concrete examples.

For gauge theories also (with no Fayet-Iliopoulos term), if supersymmetry is spontaneously broken at the tree level by the chiral superpotential and the potential has the same minimum value along a valley, higher order corrections will remove the degeneracy. However now the effective potential does not necessarily increase with the scalar fields[23] and can in some cases reach its minimum for large values of the fields. This fact has led Witten to suggest a possible "inverse" solution of the hierarchy problem, in which the small mass scale μ is put into the theory at the start and the large mass scale M is generated by radiative corrections.

In a gauge theory with a U(1) factor and no Fayet-Iliopoulos term, can one be generated in perturbation theory and cause spontaneous breaking of supersymmetry? In the one-loop approximation a D tadpole is quadratically divergent and proportional to the trace of the U(1) charge Y. It has been shown[28] by the supergraph method that all higher loop contributions cancel. Therefore, if Tr Y = 0 no Fayet-Iliopoulos term is generated. It should be possible to understand this non-renormalization result as a consequence of combined supersymmetry and gauge invariance.[29] In the so called Wess-Zumino gauge, where only the physical fields and the auxiliary field D of the vector supermultiplet remain, the D tadpole, including all radiative corrections, can be related to the D-scalar-scalar vertex, by cutting a line. This vertex, in turn, is related by supersymmetry to the vector-scalar-scalar vertex for which gauge invariance provides a non-renormalization statement.

ACKNOWLEDGMENT

This work was supported by the Director, Office of High Energy and Nuclear Physics, Division of High Energy Physics of the U.S. Department of Energy under Contract No. W-7405-ENG-48. The author is on leave from CERN, Geneva Switzerland, where the present work was initiated.

REFERENCES

1. For recent reviews see P. Langacker, Physics Reports 72, 185 (1981): John Ellis, this volume.
2. E. Gildener and S. Weinberg, Phys. Rev. D13, 3333 (1976); E. Gildener, Phys. Rev. D14, 1667 (1976); S. Weinberg, Phys. Lett. 82B, 387 (1979).
3. E. Witten, Princeton Univ. preprint (1981).
4. J. Wess and B. Zumino, Phys. Lett. 49B, 52 (1974); J. Iliopoulos and B. Zumino, Nucl. Phys. B76, 310 (1974); S. Ferrara, J. Iliopoulos and B. Zumino, Nucl. Phys. B77, 413 (1974); J. Wess and B. Zumino, Nucl. Phys. B78, 413 (1974); B. Zumino, Nucl. Phys. B89, 535 (1975).
5. K. Fujikawa and W. Lang, Nucl. Phys. B88, 61 (1975).
6. S. Ferrara and O. Piguet, Nucl. Phys. B93, 261 (1975).

7. M. T. Grisaru, W. Siegel and M. Rocek, Nucl. Phys. B159, 429 (1979).

8. T. N. Sherry, ICTP Trieste preprint IC/79/105 (1979); S. Dimopoulos and S. Raby, SLAC preprint (1981); M. Dine, W. Fischler and M. Srednicki, IAS Princeton preprint (1981); S. Dimopoulos, S. Raby and F. Wilczek, Univ. of California at Santa Barbara preprint ITP-81-31 (1981); R. K. Kaul and P. Majumdar, Center for Theor. Studies preprint, Bangalore (1981); N. Sakai, Tohoku Univ. preprint (1981); S. Dimopoulos and H. Georgi, Harvard Univ. preprint HUTP-81/A 022 (1981); S. Dimopoulos and F. Wilczek, Univ. of California at Santa Barbara preprint (1981); L. E. Ibáñez and G. G. Ross, Oxford Univ. preprint (1981); H. P. Nilles and S. Raby, SLAC-PUB-2743 (1981); S. Weinberg, Harvard Univ. preprint (1981).

9. Discussions of soft explicit breaking are given by J. Iliopoilos and B. Zumino, Nucl. Phys. B76, 310 (1974); K. Harada and N. Sakai, Tohoku Univ. preprint (1981) L. Girardello and M. Grisaru, Harvard-Brandeis preprint (1981).

10. L. O'Raifeartaigh, Phys. Lett. 56B, 41 (1975); Nucl. Phys. B96, 331 (1975).

11. P. Fayet, Phys. Lett. 58B, 67 (1975).

12. K. Fujikawa and W. Lang, Nucl. Phys. B88, 77 (1975).

13. G. Woo, Phys, Rev. D12, 975 (1975).

14. S. Weinberg, Phys. Lett. 62B, 111 (1976).

15. D. M. Capper and M. Ramón Medrano, J. Phys. G2, 269 (1976).

16. P. West, Nucl. Phys. B106, 219 (1976).

17. L. O'Raifeartaigh and G. Parravicini, Nucl. Phys. B111, 516 (1976).

18. W. Lang, Nucl. Phys, B114, 123 (1976).

19. I thank M. Grisaru for explaining to me this simple and general argument.

20. The results stated in this paragraph and the next were obtained in collaboration with V. Glaser and R. Stora.

21. This fact was also noted by H. Nicolai, private communication to R. Stora.

22. M. Huq, Phys. Rev. D14, 3548 (1976).

23. E. Witten, Phys. Lett. 105B, 267 (1981).

24. P. Fayet and J. Iliopoulos, Phys. Lett. 51B, 461 (1974).

25. S. Coleman and E. Weinberg, Phys. Rev. D7, 1888 (1973).

26. B. A. Ovrut and J. Wess, IAS Princeton preprint (1981).

27. The results of this paragraph and the next were obtained in collaboration with J. Wess.

28. W. Fischler, H. P. Nilles, J. Polchinski, S. Raby and L. Susskind, Phys. Rev. Lett. 47, 757 (1981).

29. M. K. Gaillard, unpublished.

SUPERSYMMETRIC SOLITON STATES IN EXTENDED SUPERGRAVITY THEORIES

G.W. Gibbons
D.A.M.T.P.
University of Cambridge
Silver Street
Cambridge
CB3 9EW
U.K.

It seems that the extended N = 8 supergravity theory [1] offers the best hope at present for a finite theory of gravity [2]. This theory is a unified theory in which the graviton is on the same footing and in the same supermultiplet as 8 spin $\frac{3}{2}$ Majorana gravitini, 28 spin 1 graviphotons, 56 Majorana spin $\frac{1}{2}$ particles and 70 spin 0 particles. These particles are all massless and carry no electric or magnetic charges corresponding to the graviphoton fields. At large scales — corresponding to weak coupling and the semi-classical limit one expects there to be a fairly clear cut distinction between the gravitational field — represented by a classical spacetime and the other matter fields whereas at small scales and strong coupling one would expect this clear cut distinction to break down and all particles to interact equally strongly.

How can one hope to probe this theory at small scales where quantum effects are strongest? One possibility is to use some discrete approximation to the path integral like the Regge Calculus [3] and then employ Monte Carlo techniques. Another approach is to examine the soliton structure of the theory in the hope that some sort of "duality" holds whereby the solutions which are a feature of the weak coupling limit play an important role in the strong coupling limit. This phenomenon is conjectured [3, 4, 5] to be important in the N = 4 supersymmetric Yang-Mills theory which is believed to be self dual and a similar approach has been suggested by Hajicek [6] for studying non-supersymmetric quantum gravity theories. Solitons may also be important in phenomenological considerations since the observed particles cannot be described using the fundamental fields listed above even supposing they acquire masses.

In this report I shall describe the soliton structure of the N = 0, 1, 2 and 3 supergravity theories and make some conjectures about N > 4. Before doing this I shall make some preliminary remarks about supersymmetry at finite temperature and about central charges,

1. It is clear that a Gibbs state at some non-zero temperature is not symmetrical between fermions and bosons even though the fields themselves interact supersymmetrically, because of the differences between the Bose-Einstein and Fermi-Dirac distributions. In fact the requirement that the fermion fields in the path integral be antiperiodic in imaginary time is incompatible with the supersymmetry algebra which requires the super-symmetry generators to commute with the hamiltonian [7]. This means that one cannot act on a Gibbs state with supersymmetry generators to generate supermultiplets as one can on a pure state.

2. It has been pointed out by Olive and Witten [4] that if one starts with a theory providing a realization on the fundamental fields of a supersymmetry algebra without central charges [8], the existence of solitons may force (for $N \geq 2$) the introduction of central charges which occur as boundary terms. Since central charges have the dimensions of mass they cannot be carried by massless fields and so supergravity theories realize the algebra without central charges unless one considers configurations with a non zero mass M. Then central charges can — and according to Teitleboim [9] do arise for $N \geq 2$ supergravity their values, Z^{ij}, being given by the surface integrals

$$Z^{ij} = \frac{1}{4\pi G^{\frac{1}{2}}} \int_{\infty} (F^{ij}_{\mu\nu} + \frac{i}{2}\, \varepsilon_{\mu\nu}{}^{\lambda\rho}\, F^{ij}_{\lambda\rho})\, d\Sigma^{\mu\nu} \qquad (1)$$

where $F^{ij}_{\mu\nu}$ is the graviphoton field strength and is antisymmetric on the spacetime indices $\mu\nu$ which run from 0 to 3 and also on the internal indices ij which run from 1 to N, G is Newton's constant and has dimensions mass^{-2}, h and c have been set to unity and the integral is taken over a suitable 2-sphere at infinity. Z^{ij} contains both electric and magnetic type charges. Any state must, for entirely algebraic reasons satisfy an important Bogomolny type inequality [4, 10] : its mass, M, is bounded below by the moduli of the eigenvalues of the central charge operator Z^{ij}. In terms of the electric and magnetic charges Q_n and P_n, $n = [\frac{N}{2}]$, [ ] denotes integer part,

$$G^{\frac{1}{2}} M \geq (Q_n^2 + P_n^2)^{\frac{1}{2}} \qquad (2)$$

Since fundamental fields cannot carry central charges (if massless (2) shows that states satisfying the equality may decay by emission of massless quanta so reducing their mass but since the central charges cannot be emitted states satisfying the equality in (2) should be stable.

A general massive supermultiplet with central charges has 2^{2N} states with a spin range of $\frac{N}{2}$ compared with the massless multiples which have 2^N states with a spin range of $\frac{N}{4}$. If one or more bounds in (2) are satisfied however the number of states in the multiplets are reduced and for even N if 2 holds for $1 \leq n \leq \frac{N}{2}$ the multiplets are identical in structure to the corresponding massless multiplets with the same N. [10]. For odd N the spin range is as for N+1, if (2) holds for $1 \leq n \leq \frac{N-1}{2}$. This reduction in the multiplet size arises because some of the supersymmetry generators annihilate the state. They all annihilate the vacuum state which is completely invariant under supersymmetry. States satisfying the Bogomolny bound retain some but not all of this invariance.

Armed with this information we turn to the question of soliton states in supergravity theories. In Einstein's theory (N=0) there are no stable solitons [6] since the Hawking thermal evaporation [11] makes the only natural candidates — black holes of the Kerr family quantum mechanically unstable against the emission of gravitons. Note that even the maximally rotating Kerr solutions which have zero temperature are unstable against superradiant loss of angular momentum [12, 13].

In N=1 supergravity the situation is the same. The only candidate solutions are black holes of the Kerr - family. No hair theorems indicate (but do not entirely prove) that no other solutions — non singular outside an event horizon exist unless the Rarita-Schwinger field ψ_μ is "pure gauge" — i.e. of the form $\psi_\mu = \nabla_\mu \varepsilon$ where ε is a spinor field [14, 15, 16]. These black holes are even more unstable since emission of gravitini is also possible and will proceed at a faster rate than for gravitons since the spin-dependent centrifugal barrier will be lower. It has been claimed [17, 14] that one can construct "super translated" superpartners of Schwarzschild black holes using spinors ε which become constant at infinity and satisfying some gauge condition — for instance being solutions of the massless Dirac equation [17]. This is not correct, the proposed Rarita-Schwinger fields are in fact singular on the horizon. This is because non-singularity on the horizon for a spinorial field requires antiperiodicity in imaginary time with the thermal period $\beta = T^{-1} = 8\pi M$ [18] which contradicts the required constancy of the field at infinity. We encounter here precisely the conflict between supersymmetry and finite temperature mentioned above and discussed in detail in [7]. Thermal black hole states are not supersymmetric — for instance the emission rates for gravitons and for gravitini will differ.

For N = 2 the situation is different. This is the first case in which central charges enter. Candidate soliton solutions are the electrically and magnetically charged Kerr-Newman family. Since Massless fields can carry angular momentum the Hawking emission will reduce the angular momentum to zero, reducing the mass but leaving unaltered the central electric, and magnetic charges Q and P respectively. The mass must however satisfy for purely classical reasons unrelated to supersymmetry — the inequality 2) if no naked singularity is to develop. Indeed 2) is closely related to Penrose's Cosmic Censorship hypothesis [19] and in some sense a consequence of it [20]. The temperature of a black hole of mass M with electric and magnetic charges is

$$T = G^{-1}(M^2 - Z^2)^{\frac{1}{2}}/2\pi(M^2 + (M^4 - Z^2M^2)^{\frac{1}{2}}) \qquad (3)$$

where

$$GZ^2 = Q^2 + P^2 \qquad (4)$$

The temperature vanishes in the limit when the inequality is satisfied. Such extreme black holes are quantum mechanically stable since no massless quanta are radiated at zero temperature. It is essential for this stability that the fundamental fields are electrically and magnetically neutral — i.e. that the 0(2) symmetry is not gauged. For the same reason neither Q nor P is quantized (at least classically) — there is no unit of charge in the theory. Despite this fact one can think of Q and P as "topological". There is no local density of electric or magnetic charge, neither is there a "topological current" but the topology of spacetime is non-trivial ($R^2 \times S^2$ with Euler number 2). This is the phenomenon called by Wheeler [21] "charge without charge". One way of quantizing the charge might be to demand that the entropy of these solutions πG^2M^2, equal the logarithm of the number of states in the super-multiplet. This gives $\pi G M^2 = \text{Ln } 4$ for $N = 2$.

In order to fit these solitons into supermultiplets it is necessary to examine the "zero modes" that arise when computing quantum corrections. In the 0(2) theory a pure gauge complex Rarita-Schwinger field χ_μ is of the form [22]

$$\chi_\mu = \hat{D}_\mu \chi \qquad (5)$$

where $\hat{D}_\mu$ is the supercovariant derivative and acting on the complex spinor χ has the form (expanding at linearized level around an Einstein-Maxwell background).

$$\hat{D}_\mu \chi = \nabla_\mu \chi - \frac{1}{4} F_{\alpha\beta} \gamma^\alpha \gamma^\beta \gamma_\mu \chi \qquad (6)$$

If one imposes the gauge condition $\gamma^\mu \chi_\mu = 0$ one has fixed the gauge freedom (5) except for that arising from Dirac fields χ satisfying $\gamma^\mu \hat{D}_\mu \chi = \gamma^\mu \nabla_\mu \chi = 0$ which give rise to square integrable Rarita-Schwinger fields but which themselves are not square integrable. Those χ which are constant at infinity generate the supertranslation zero modes — this is just the fermionic equivalent of the familiar black hole translation modes [17, 23]. As explained above such non-singular solutions can exist only for black holes at zero temperature. Explicit computation shows that in the extreme Reisner-Nordstrom solutions for each Q and P attaining the bound (2) 4 such asymptotically constant Dirac fields exist. They are time independent and the spin can point up or down. Two of these solutions satisfy $\hat{D}_\mu \chi = 0$ and these are true supersymmetries: the classical solution is invariant under supersymmetry transformation generated by these spinor fields. The other two can be used to generate superpartners of the original spin 0 state, with spins $\pm\frac{1}{2}$ and 0. Thus each extreme black hole can be fitted into the basic 0(2) multiplet with maximal central charge. The states themselves carry an extra Usp(1) invariance. One has in addition an invariance which enables one to rotate the electric and magnetic fields into themselves. Presumably the relation $GM^2 = Q^2 + P^2$ will remain true to all orders in h just as is believed to hold in Yang-Mills theory.

In the 0(3) theory [4] things are similar. The extreme Reisner-Nordstrom solutions are also 0(3) solutions. For fixed electric and magnetic charges there are 6 solutions of the Dirac equation which are constant at infinity, 2 of which are supercovariantly constant. One can use the remaining 4 to generate states of spin 0, $\pm\frac{1}{2}$, ± 1 which transform under Usp(2).

In addition to these single-soliton solutions there are static multi-soliton solutions which exhibit "antigravity" [27]. They describe arbitrary numbers of extreme black holes placed at arbitrary locations [28, 29, 30]. In 0(2) there are 4 asymptotical constant spinors two of which are supercovariantly constant.

I have not completed my analysis of the N=4 theory. In the SU(4) version [25] it appears that the solitons will fit into N=4 multiplets with spins up to spin 1. However this is if the scalar fields introduce no extra solitons beyond those discussed above. If this pattern persists up to N=8 one would have massive solitons up to spin 2 which is perhaps a hint of self-duality.

In the N=2 case one has a candidate effective field theory — the 0(2) multiplet with central charge may be coupled to 0(2) supergravity as shown by Zachos [26]. This model has a potential for the scalars which is not bounded below which may indicate that the 0(2) theory is unstable against the formation of a condensate of solitons.

Further details of this work will appear elsewhere. I thank Alessandra D'Adda, P. Hajicek, S.W. Hawking, C. Hull, M. Rocek and A. Yuille for many helpful discussions.

[1] E. Gremmer and B. Julia, Phys. Letts. 80B 48 (1978).

[2] S.M. Christensen, M.J. Duff, G.W. Gibbons and M. Rocek, Phys. Rev. Letts. 45 161 (1980).

[3] C. Montonen and D. Olive, Phys. Letts. 72B 213 (1977).

[4] D. Olive and E. Witten, Phys. Letts. 78B 97 (1978).

[5] H. Osborn, Phys. Letts. 83B 321 (1979).

[6] P. Hajicek, "Quantum Wormholes" I and II Bern preprints to appear in Nucl. Phys. B.

[7] L. Girardello, M.T. Grisaru and P. Salomonson, Nucl. Phys. B178 331 (1981).

[8] R. Haag, J.T. Loupszanski and M. Sohnuis, Nucl. Phys. B88 257 (1975).

[9] C. Teitleboim, Phys. Letts. 69B 240 (1977).

[10] S. Ferrara, C.A. Savoy and B. Zumino, Phys. Letts. 100B 393-398 (1981).

[11] S.W. Hawking, Nature 248 30 (1974).

[12] W. Unruh, Phys. Rev. D10 3194 (1974).

[13] For reviews and references on black hole thermodynamics see

G.W. Gibbons in S.W. Hawking and W. Israel, "General Relativity", Cambridge University Press 1979.

G.W. Gibbons, LPTNS preprint 80/28.

[14] P. Cordero and C. Teitleboim, Phys. Letts. 78B 80 (1978).

[15] I.F. Urrutia, Phys. Letts. 89B 52 (1979).

[16] R. Güven, Phys. Rev. D22 2327 (1980).

[17] T. Yoneya, Phys. Rev. D17 2567 (1978).

[18] G.W. Gibbons and M.J. Perry, Proc. Roy. Soc. A358 467 (1978).

[19] R. Penrose, Ann. N.Y. Acad. Sci. 224 125 (1973).

[20] P.S. Jang, Phys. Rev. D20 834 (1979).

[21] J.A. Wheeler, "Geometrodynamics", Academic Press New York (1962).

[22] S. Ferrara and P. van Niewenhuizen, Phys. Rev. Letts. 37 1669 (1976).

[23] G.W. Gibbons and M.J. Perry, Nucl. Phys. B146 90 (1978).

[24] D.Z. Freedman, Phys. Rev. Letts. 38 105 (1977).

[25] E. Gremmer, J. Scherk and S. Ferrara, Phys. Letts. 74B 61 (1978).

[26] K. Zachos, Phys. Letts. 76B 329 (1978).

[27] J. Scherk, Phys. Letts. 88B 265 (1979).

[28] A. Papapetrou, Proc. Roy. Irish Acad. A51 191 (1947).

[29] S.D. Majundar, Phys. Rev. 72 390 (1947).

[30] J.B. Hartle and S.W. Hawking, Commun. Math. Phys. 26 87 (1972).

STABILITY PROPERTIES OF GRAVITY THEORIES

S. Deser *)
CERN, Theory Division
1211 Geneva 23, Switzerland

ABSTRACT

We study the stability properties of general relativity with a non-vanishing cosmological constant Λ by means of the energy. First, it is shown that there exists a suitable definition of energy in these models, for all metrics tending asymptotically to any background solution which has a timelike Killing symmetry. It is conserved and has flux integral form. Stability is established for all systems tending asymptotically to anti-De Sitter space when $\Lambda < 0$, using supergravity techniques. Spinorial charges are defined which are also flux integrals and satisfy the global graded anti-De Sitter algebra. The latter then implies that the energy is always positive. For $\Lambda > 0$, it is shown that small excitations about De Sitter space are stable, provided they occur within the event horizon intrinsic to this space. Outside the horizon an instability arises which signals the onset of Hawking radiation; it is shown to be universal to all systems. Semi-classical stability is also discussed for $\Lambda > 0$.

*) Supported in part by NSF grant PHY-78-09644 A01.

1. INTRODUCTION

One of the open problems in current physics is the observed smallness of the cosmological constant Λ, or equivalently of the vacuum energy density of the Universe. From the point of view of particle physics, this is highly unnatural, requiring extreme fine tuning of parameters. One possible way to exclude the cosmological constant, already at the classical level, would be to show that it leads to some fundamental instabilities in the Einstein theory. This was the motivation for the present study, which has been carried out in collaboration with L.F. Abbott at CERN; details may be found in a forthcoming joint paper (Nucl. Phys. B).

Stability of a bounded matter system in flat space is usually established by showing it to have positive energy, with respect to a lowest, vacuum, state. For gravity, (with $\Lambda = 0$) energy of any asymptotically flat solution is also perfectly definable with respect to flat space as vacuum. It turns out that this energy is always positive and that the theory (also in the presence of positive energy matter) is stable; quite general and rigorous results have been obtained in recent years[1)-4)]. On the other hand, when $\Lambda \neq 0$, flat space is no longer an acceptable background (since it does not solve the Einstein equations), but must be replaced as vacuum by the "flattest", maximally symmetric solutions of the cosmological equations, namely De Sitter O(4,1) or anti-De Sitter O(3,2) space according to whether $\Lambda > 0$ or $\Lambda < 0$, respectively. At this point, a number of problems arise for the stability programme. First, can any reasonable physical substitute for energy be defined ? Having lost asymptotic Poincaré invariance, one is left with the asymptotic De Sitter or anti-De Sitter algebra at infinity, for which P_μ^2 is no longer a Casimir operator, being replaced by the five-dimensional "rotations" $J_{ab} = -J_{ba}$ $(a,b = 0,\ldots,4)$. One must therefore show that J_{04}, which becomes P_0 upon contraction $(\Lambda \to 0)$, is acceptable and that this quantity is really definable as a flux integral at infinity so as to satisfy the asymptotic global algebra. This is accomplished in Section 2, through a general analysis of how to obtain and interpret conserved quantities with respect to a background which possesses symmetries. We will see that, associated to every such symmetry, there is a generator which is conserved and of flux integral form. The preferred one, which we call the Killing energy, is that which is connected with a timelike Killing vector or symmetry. Fortunately, for both signs of Λ, the maximally symmetric backgrounds have timelike symmetries, associated with J_{04}. The other nine generators are not timelike (nor are $(P_i, J_{\mu\nu})$ when $\Lambda = 0$). A brief review of the properties of De Sitter spaces in Section 3 will show that for $\Lambda > 0$, the necessary presence of an event horizon, where the timelike Killing vector becomes null and then spacelike requires a more careful analysis of stability. Within the horizon, however, and in all space for $\Lambda < 0$ (where there is no horizon) it will be seen in Section 4 that small oscillations about vacuum have positive energy. The possibility of negative energy for excitations outside the horizon (for $\Lambda > 0$) is a reflection, in Hamiltonian form, of the generic features which lead to Hawking radiation[5)]. Next we shall discuss semi-classical stability, against quantum mechnical tunnelling, for $\Lambda > 0$, which would be violated in the presence of Euclidean "bounce" solutions[6)-8)]; no evidence for these is found. The following Section, 6, is devoted to a demonstration of stability for all asymptotically anti-De Sitter solutions when $\Lambda < 0$ by showing that the full energy is positive in that case. Here one uses methods of supergravity, parallel to those which were used to establish[2)] positivity of the energy for $\Lambda = 0$. To accomplish this it is first necessary to define spinorial charges which are conserved and also have flux integral form and show the corresponding existence of appropriate Killing spinors, whose presence is implicit but rather trivial when $\Lambda = 0$.

The outcome of our analysis is thus the opposite of what had been hoped for, and we end up by praising, rather than burying, theories with $\Lambda \neq 0$, at least from stability considerations (spaces with $\Lambda \neq 0$ do have pecularities[9)], which we mention, but are not directly concerned with here). On the other hand, there emerges a rather coherent picture of the quantities basic to understanding Einstein theory and their properties, whatever the value of Λ, as well as a simple Hamiltonian way of understanding the effect of event horizons on stability.

2. CONSERVED QUANTITIES

Consider the physical system defined by the Einstein equations

$$G_{\mu\nu} + \Lambda g_{\mu\nu} = 0 \tag{2.1}$$

together with a background metric $\bar{g}_{\mu\nu}$ which satisfies (2.1); we decompose the full metric $g_{\mu\nu}$ according to

$$g_{\mu\nu} = \bar{g}_{\mu\nu} + h_{\mu\nu} \tag{2.2}$$

where $h_{\mu\nu}$ is not necessarily small, but does obey the boundary condition that it vanishes asymptotically at some appropriate speed. We will construct conserved quantities from $(\bar{g}_{\mu\nu}, h_{\mu\nu})$ corresponding to the symmetries of the background. Although we are primarily concerned with background De Sitter or anti-De Sitter spaces, the method is completely general, and leads to flux integral expressions for those generators, which are constructed from the gravitational stress-tensor and the appropriate Killing vectors. Our conventions are that $R_{\mu\nu} \equiv R^{\alpha}{}_{\mu\alpha\nu} \sim + \partial_{\alpha}\Gamma^{\alpha}_{\mu\nu}$, signature (+++-) and all operations such as covariant differentiation $(\bar{D}_{\mu})$ or index moving are with respect to $\bar{g}_{\mu\nu}$. We define the symmetric stress tensor $T^{\mu\nu}$ to be all terms of second and higher order in $h_{\mu\nu}$ when the decomposition (2.2) is inserted in (2.1):

$$G_L^{\mu\nu}(\bar{g}, h) + \Lambda h_{\mu\nu} = T^{\mu\nu} = T^{\nu\mu}. \tag{2.3}$$

The subscript L refers to terms linear in $h_{\mu\nu}$. Using the fact that $G_{\mu\nu}(\bar{g}) + \Lambda\bar{g}_{\mu\nu} = 0$ and that the left side of (2.3) therefore obeys the (exact) linearized Bianchi identity $\bar{D}_{\mu}(G_L^{\mu\nu} + \Lambda h^{\mu\nu}) = 0$, the field equations imply covariant conservation of $T^{\mu\nu}$:

$$\bar{D}_{\mu} T^{\mu\nu} = 0. \tag{2.4}$$

To turn covariant into ordinary conservation, we have to define a conserved (background) contravariant vector density from the tensor $T^{\mu\nu}$ since then, and only then, is $\bar{D}_{\mu} J^{\mu} \equiv \partial_{\mu} J^{\mu}$. When $\bar{g}_{\mu\nu}$ has a symmetry, there exists a Killing vector ξ_{μ}, obeying

$$\bar{D}_{\mu} \bar{\xi}_{\nu} + \bar{D}_{\nu} \bar{\xi}_{\mu} = 0. \tag{2.5}$$

Consequently,

$$\bar{D}_{\mu}(T^{\mu\nu} \bar{\xi}_{\nu}) = (\bar{D}_{\mu} T^{\mu\nu}) \bar{\xi}_{\nu} + \tfrac{1}{2} T^{\mu\nu}(\bar{D}_{\mu}\bar{\xi}_{\nu} + \bar{D}_{\nu}\bar{\xi}_{\mu}) = 0 \tag{2.6}$$

and $\sqrt{-\bar{g}}T^{\mu\nu}\bar{\xi}_\nu$ is the desired contravariant tensor density, for which true conservation holds:

$$\bar{D}_\mu(\sqrt{-\bar{g}}\, T^{\mu\nu}\bar{\xi}_\nu) \equiv \partial_\mu(\sqrt{-\bar{g}}\, T^{\mu\nu}\bar{\xi}_\nu) = 0. \tag{2.7}$$

So to every Killing vector, there is associated a conserved generator

$$E(\bar{\xi}) = 1/8\pi G \int d^3x \sqrt{-\bar{g}}\, T^{0\nu}\bar{\xi}_\nu \tag{2.8}$$

In particular, if $\bar{\xi}_\nu$ is timelike, this is the Killing energy. Despite the fact that $\Lambda \neq 0$, we now show that $E(\bar{\xi})$ can be written in flux integral form, just as for $\Lambda = 0$. From (2.3) it follows that

$$T^{\mu\nu} \equiv \bar{D}_\alpha \bar{D}_\beta K^{\mu\alpha\nu\beta} + \tfrac{1}{2}(R^\mu{}_{\alpha\beta}{}^\nu H^{\alpha\beta} - \Lambda H^{\mu\nu}) \tag{2.9}$$

where the superpotential K is defined by

$$\begin{aligned} 2K^{\mu\alpha\nu\beta} &\equiv \bar{g}^{\mu\beta}H^{\nu\alpha} + \bar{g}^{\nu\alpha}H^{\mu\beta} - \bar{g}^{\mu\nu}H^{\alpha\beta} - \bar{g}^{\alpha\beta}H^{\mu\nu}, \\ H^{\mu\nu} &\equiv h^{\mu\nu} - \tfrac{1}{2}\bar{g}^{\mu\nu}h_\alpha{}^\alpha . \end{aligned} \tag{2.10}$$

It has the algebraic symmetries of the Riemann tensor:

$$K^{\mu\alpha\nu\beta} = -K^{\alpha\mu\nu\beta} = K^{\alpha\mu\beta\nu} = K^{\nu\beta\mu\alpha}. \tag{2.11}$$

Symmetry and conservation of $T^{\mu\nu}$ in (2.9) can easily be checked using the background field equations and its derivative consequences,

$$\bar{D}_\beta \bar{R}_{\mu\nu\alpha}{}^\beta \equiv \bar{D}_\nu \bar{R}_{\alpha\mu} - \bar{D}_\mu \bar{R}_{\alpha\nu} = 0, \quad \bar{D}_\beta \bar{R}^{\mu\nu} = 0$$

but without any assumptions on the full Riemann tensor.

Next, we form $T^{\mu\nu}\bar{\xi}_\nu$, and recast it into the expression

$$\begin{aligned} \sqrt{-\bar{g}}\, T^{\mu\nu}\bar{\xi}_\nu &= \sqrt{-\bar{g}}\, \bar{D}_\alpha[(\bar{D}_\beta K^{\mu\alpha\nu\beta})\bar{\xi}_\nu - K^{\mu\beta\nu\alpha}\bar{D}_\beta\bar{\xi}_\nu] \\ &\equiv \bar{D}_\alpha \mathcal{F}^{\mu\alpha}. \end{aligned} \tag{2.12}$$

It may be checked that all additional terms "miraculously" vanish upon use of the Killing identity $\bar{D}_\beta \bar{D}_\alpha \bar{\xi}_\nu + R^\lambda{}_{\beta\alpha}{}^\nu \bar{\xi}_\nu \equiv 0$. Furthermore, the quantity $^{\mu\alpha}$ is (almost obviously) an antisymmetric tensor density, and therefore its divergence is an ordinary one, $\bar{D}_\alpha \mathcal{F}^{\mu\alpha} \equiv \partial_\alpha \mathcal{F}^{\mu\alpha}$. Hence the desired result:

$$8\pi G\, E(\bar{\xi}) = \int d^3x \sqrt{-\bar{g}}\, T^{0\nu} \bar{\xi}_\nu = \oint dS_i\, \mathcal{F}^{0i} \tag{2.13}$$

As a check, when $\Lambda = 0$ and the background is chosen to be flat, introduction of Cartesian co-ordinates simplifies (2.13) to be the usual expression for the Poincaré generators. In particular, when $\bar{\xi}_\mu = (1, \vec{0})$ we obtain the standard energy formula

$$16\pi G\, E = \oint dS_i \left(h_{ij,j} - h_{jj,i} \right) \tag{2.14}$$

which correctly reproduces the mass of any asymptotically Schwarzschild metric. Note incidentally that the ten Poincaré generators automatically obey the (global) Poincaré algebra simply by virtue of the definition (2.13) and the algebra of the ten Poincaré Killing vectors (A = 1-10).

When $\Lambda \neq 0$, with $\bar{g}_{\mu\nu}$ a De Sitter or anti-De Sitter metric, we would obtain the 10 Killing generators corresponding to the background De Sitter or anti-De Sitter symmetries, and they also automatically satisfy the appropriate global algebra. In particular, we get the timelike $E(\bar{\xi})$ expression[*)]. We also mention that this whole procedure could also have been carried out in first order form (used in Section 4) to yield $E(\bar{\xi})$ there as well.

We complete this section with a treatment of the graded algebra which can be introduced when $\Lambda < 0$ (but not $\Lambda > 0$!), in terms of spinorial charges Q. The resulting local supersymmetry is that of supergravity with a cosmological term[10)] and a spin 3/2 "mass" term[11)]. This will be used in Section 6 to show that the supergravity energy operator is positive and from this establish stability for classical gravity with $\Lambda < 0$. First, however, we must show that the spinor charges can be written as surface integrals as in (2.13) so as to satisfy the graded global algebra at infinity. The spinorial charge density is

$$Q^\mu = \varepsilon^{\mu\alpha\beta\nu} \gamma_5 \bar{\gamma}_\alpha \bar{D}_\beta \psi_\nu \, . \tag{2.15}$$

Its origin may be understood, just like that of $T^{\mu\nu}$, in terms of a decomposition of the full Rarita-Schwinger equation into a "linear" part and a remainder. Here ψ_ν

---

*) A check here is to verify that the equivalent of the Schwarzschild solution for $\Lambda \neq 0$ has energy m. For $\Lambda > 0$, there are corrections because one must stay within the unavoidable event horizon (rather than go to infinity) in calculating the flux integral; apart from these, the correct result E = m emerges.

is the spin 3/2 field, $\bar{\gamma}_\alpha$ are the background covariant γ matrices with respect to the background vierbein and the modified covariant derivative on a spinor, $\tilde{D}_\beta$, defined by

$$\tilde{D}_\beta \equiv \bar{D}_\beta + \tfrac{1}{2} m \bar{\gamma}_\beta \quad , \quad m^2 \equiv \tfrac{1}{3} |\Lambda| \tag{2.16}$$

has the basic property that $[\tilde{D}_\beta, \tilde{D}_\alpha] = 0$ for a background anti-De Sitter space. The current Q^μ satisfies $\tilde{D}_\mu Q^\mu = 0$, and to convert this to an ordinary conservation law, we introduce Killing spinors obeying $\tilde{D}_\mu \alpha = 0$ (consistent with $[\tilde{D}_\nu, \tilde{D}_\mu]\alpha = 0$). The quantity $\bar{\alpha} Q^\mu$ is easily seen to take the form

$$\bar{\alpha} Q^\mu = \bar{D}_\beta (\bar{\alpha}\, \varepsilon^{\mu\alpha\beta\nu} \gamma_5 \bar{\gamma}_\alpha \psi_\nu) = \partial_\beta (\bar{\alpha}\, \varepsilon^{\mu\alpha\beta\nu} \gamma_5 \bar{\gamma}_\alpha \psi_\nu). \tag{2.17}$$

The last equality follows because the quantity in parentheses is an antisymmetric tensor density. But since $\bar{\alpha} Q^\mu$ is a contravariant vector, we see immediately that its ordinary divergence vanishes identically, $\partial_\mu (\bar{\alpha} Q^\mu) \equiv 0$, so that the spinor charge is both conserved and has the flux form

$$Q(\alpha) \equiv \int d^3x\, \bar{\alpha} Q^0 = \oint dS_j\, \bar{\alpha}\, \varepsilon^{0ijk} \gamma_5 \bar{\gamma}_i \psi_k). \tag{2.18}$$

This is the required analogue of (2.13) for the bosonic generators. [The analogy actually goes further, in that only one "Coulomb" component of ψ_k enters in (2.18), corresponding to the "Coulomb" part of the metric in (the appropriate generalization of) Eq. (2.14)].

Each of the four independent Killing spinors $\alpha^{(\beta)}_{\beta'}$, where β' is the spinor index and (β) is the label of each spinor, defines a fermionic charge $Q_{(\beta)}$. These then satisfy the graded relation

$$\{Q_{(\beta)}, \bar{Q}_{(\beta')}\} = \tfrac{1}{2} (\gamma^{(\mu)})_{(\beta\beta')} J_{(\mu 4)} + (\sigma^{(\mu\nu)})_{(\beta\beta')} J_{(\mu\nu)} \tag{2.19}$$

if we take the α's to commute (for convenience). Having defined the required conserved quantities, we turn next to the choice of Killing vectors.

3. DE SITTER SPACES

We give a brief review of the symmetries of "vacuum" spaces when $\Lambda \neq 0$. De Sitter space corresponds to a four-surface $z_\mu^2 + z_4^2 = 3/\Lambda$, $\Lambda > 0$, in flat five-space. Among the rotations of the embedding space are the boosts mixing z_0 with (z_i, z_4). For example $\xi_a = (z_4, 0, 0, 0, z_0)$ is a timelike Killing vector when $|z_4| > |z_0|$, but signals the existence of an event horizon at $|z_4| = |z_0|$, where stability must be discussed separately, since $E(\bar{\xi})$ no longer acts like an energy beyond it. Of course, an observer will only interact with events inside the horizon, which means that $E(\bar{\xi})$ tests stability to excitations visible to the observer. It is illuminating to apply these ideas to a simple model, namely a scalar field in De Sitter space. Representing the metric in the form

$$ds^2 = -dt^2 + f^2(t)(dx^2+dy^2+dz^2)\,, \quad f(t) \equiv \exp\sqrt{\tfrac{\Lambda}{3}}\,t \tag{3.1}$$

with the "timelike" vector

$$\bar{\xi}^\mu = \left(-1, \sqrt{\tfrac{\Lambda}{3}}\,\vec{x}\right)\,, \quad \bar{\xi}^2 = -1 + \tfrac{\Lambda}{3}\,|\vec{x}f|^2 \tag{3.2}$$

we see that the horizon appears (at any given time) for distances such that $|\sqrt{\Lambda/3}\,\vec{x}f| = 1$. The action and energy momentum densities for this theory are

$$\begin{aligned} I &= \int d^4x\, f^3\left[\tfrac{1}{2}\dot{\phi}^2 - \tfrac{1}{2}f^{-2}(\nabla\phi)^2 - V(\phi)\right] \\ -\mathcal{T}^0_{\ 0} &= \tfrac{1}{2}f^{-3}\pi^2 + \tfrac{1}{2}f(\nabla\phi)^2 + f^3V(\phi) \\ -\mathcal{T}^0_{\ i} &= \pi\,\partial_i\phi \end{aligned} \tag{3.3}$$

where $\pi \equiv f^3\dot{\phi}$. The energy density is positive [if $V(\phi)$ is] but time-dependent. The conserved energy (which is of course not a flux integral here) is still $E(\bar{\xi}) = \int d^3x\, \mathcal{T}^{0\mu}\xi_\mu$. The integrand has the form

$$\mathcal{T}^{0\mu}\bar{\xi}_\mu = \left\{\tfrac{1}{2}\left[f^{-3}\pi^2 + f(\nabla\phi)^2\right] - \sqrt{\tfrac{\Lambda}{3}}\,\vec{x}\cdot\pi\vec{\nabla}\phi\right\} + f^3V(\phi). \tag{3.4}$$

It will be positive provided the bracketed quantity is. The triangle inequality

$$\tfrac{1}{2}(\vec{A}^2 + \vec{B}^2) > \sqrt{\tfrac{\Lambda}{3}}\,f\,|\vec{x}|\,\vec{A}\cdot\vec{B}\,, \quad \vec{A} = f^{-3/2}\pi\hat{x}\,,\ \vec{B} = f^{1/2}\vec{\nabla}\phi \tag{3.5}$$

makes it easy to see that the positivity condition corresponds to $\xi^2 < 0$ in (3.2), i.e., to excitations within the horizon. This correlation between the event horizon and positivity is an expression in Hamiltonian form of Hawking radiation[5], and will be seen later to be universal.

Anti-De Sitter space is the covering space for the surface $z_\mu^2 - z_4^2 = 3/\Lambda$, $\Lambda < 0$. Here there is a global timelike Killing vector corresponding to (z_0, z_4) rotation. It is $\xi_a = (z_4, 0, 0, 0, -z_0)$, $\xi^2 = -(z_4^2 + z_0^2) < 0$, since $z_4 = z_0 = 0$ is excluded. There is one peculiar feature of anti-De Sitter space which should be mentioned. Specification of initial data on a complete spacelike surface does not lead to a unique prediction of the future state of a system (including gravity itself). Radiation, not specified by the initial conditions can propagate in from infinity at a later time. Unlike the usual case where initial boundary conditions exclude incoming radiation thereafter, one is "safe" here only within ever more restricted regions of space at later times. Therefore, although the initial energy is perfectly well defined by the initial data, one can only extend the integration volume to all space at a later time if further (timelike) boundary conditions at infinity are imposed. It is in this sense that our stability results are to be understood, although the proof that energy is positive holds formally on any complete initial surface with no incoming radiation.

4. SMALL EXCITATIONS

We now apply the Killing energy together with the Killing vectors defined in the last two sections to discuss small oscillations about De Sitter or anti-De Sitter backgrounds. For this purpose, a canonical approach[12)] is most useful to discuss the $\mathcal{O}(h^2)$ part of $E(\bar{\xi})$. Indeed T^0_ν was derived to this order long ago by Nariai and Kimura[13)]. We will skip all details here, only noting that the excitations can be parametrized by the h_{ij} and their conjugate momenta p^{ij}, both being transverse traceless with respect to $\bar{\bar{g}}$.

For $\Lambda > 0$, the Hamiltonian density can be cast into the form

$$\begin{aligned} -\mathcal{T}^0{}_0 &= \tfrac{1}{2}\left(f^{-3}(P^{ij})^2 + f(\nabla Q_{ij})^2 \right) \\ -\mathcal{T}^0{}_i &= P^{jk}\,\partial_i Q_{jk} \end{aligned} \tag{4.1}$$

exactly as for the scalar field of (3.2), in terms of a canonically transformed set (Q,P). Not surprisingly, all the other features of the scalar model follow as well: $\int d^3x\, \mathcal{T}^{0\nu}\bar{\xi}_\nu$ is conserved and is positive within the event horizon $\bar{\xi}^2 < 0$. However, outside, when $\bar{\xi}^2 > 0$, the energy is no longer positive, because the triangle inequality no longer applies, as with (3.5). One would expect all physical systems to behave in this way: the free part of the energy is always $\sim\frac{1}{2}\int\{\pi^2+(\nabla\phi)^2\}$ while the momentum density is $\sim\pi\nabla\phi$. For physical matter, the non-linear parts of the energy are, like $V(\phi)$ in the scalar case, positive so the critical condition arises primarily at the free field level, where excitations beyond the horizon can give negative contributions. In particular, if the higher terms in $T^0_0(h)$ are effectively positive (as in the $\Lambda = 0$ case), then the only De Sitter instability would be that due to the horizon.

For $\Lambda < 0$, the small excitations are straightforwardly treated; this time only T^o_o is required, since we can pick a co-ordinate system which is static and in which $\bar{\xi}_i =$ = 0, and there is no horizon. The energy density is positive,

$$\mathcal{T}^0{}_0 \sim \left[\pi_{ij}^2 + \tfrac{1}{4}(\nabla h_{ij})^2 + |\tfrac{\Lambda}{6}|\, h_{ij}^2 \right], \tag{4.2}$$

and the system is stable. The masslike term in (4.2) is an artifact just like that of the spin 3/2 field[11)] in cosmological supergravity; the gravitons have only two degrees of freedom since h_{ij} is transverse-traceless.

5. SEMI-CLASSICAL STABILITY FOR $\Lambda > 0$

Having shown that $\Lambda > 0$ solutions are stable to small fluctuations about the vacuum, at least within the horizon, one may make the further test of semi-classical stability in this case, i.e., look for Euclidean "bounce" solutions whose presence would signal quantum tunnelling instability. Of course even better would be proof that the total energy is positive, which will be given for $\Lambda < 0$ in the next section; lacking this for $\Lambda > 0$, we make some comments on bounces there. In general, unusual topologies can give stability problems in gravity[14)] and semi-classical instability has been found in other gravitational contexts[7),8)].

A bounce solution here would be a metric which is asymptotically De Sitter and solves the Euclideanized Einstein equations. For example, the Euclidean continuation of the Schwarzschild-De Sitter metric would be a candidate. However, it is impossible to remove both the Schwarzschild and horizon singularities of this metric by the usual periodicity trick[15)]. In terms of Hawking radiation, De Sitter space contains radiation at a temperature fixed by the value of Λ. If a black hole could form in this space with an intrinsic temperature less than this, it would grow forever by accretion. However, for the Schwarzschild-De Sitter black hole, the black hole temperature is always larger[5)] than that of the exterior and the space is stable against this catastrophe.

Although it is doubtful on general grounds that bounce solutions exist for $\Lambda > 0$, it would clearly be desirable to extend the proofs of their absence[4),16)] for $\Lambda = 0$ to this domain as well.

6. STABILITY FOR $\Lambda < 0$

We now show that the Killing energy is positive for all excitations about the anti-De Sitter vacuum which vanish at infinity, and thereby establish stability in the $\Lambda < 0$ sector. We have already noted in Section 2 that all the generators of the graded anti-De Sitter algebra in supergravity are expressed as flux integrals, with the result that they obey the global algebra relations, in particular that of Eq. (2.19):

$$\{Q_{(\beta)}, \bar{Q}_{(\beta')}\} = \tfrac{1}{2}\, \gamma^{(\mu)}_{(\beta\beta')}\, J_{(\mu 4)} + \sigma^{(\mu\nu)}_{(\beta\beta')}\, J_{(\mu\nu)} .$$

We emphasize that in this expression, all indices are labels of particular Killing vectors or spinors. The explicit relations between the two are quite analogous to those holding in the Poincaré case, and indeed can be essentially reduced to it because the $\tilde{D}_\mu$ commute; there exists[17] a transformation $\alpha = S\eta$ which reduces the equation $\tilde{D}_\mu \alpha = 0$ to $\partial_\mu \eta = 0$. A basis for the latter is given by, e.g., $\eta_\beta = \delta_\beta^{(\beta')}$. In any case, we may now simply treat the spinor "labels" (β), (β') in (2.19), which refer to the particular Killing spinor defining the corresponding charge $Q_{(\beta)}$, as normal flat space spinor indices. Multiplying (2.19) by the numerical matrix $\gamma^{(o)}_{\beta\beta'}$ and tracing gives the positivity relation for the operator $J_{(04)}$:

$$J_{(04)} = \sum_{\beta=1}^{4} Q_{(\beta)} Q_{(\beta)} \geq 0 \tag{6.1}$$

since the $Q_{(\beta)}$ are real Majorana spinor operators. Now we just proceed as in the $\Lambda = 0$ case[2], taking matrix elements of (6.1) with no on-shell fermions and go to the tree limit, $\hbar \to 0$. This implies that $E(\bar{\xi})$, which is just this limit of $J_{(04)}$, is positive for classical $\Lambda < 0$ gravity.

We also believe, although we have not carried out the details, that Witten's recent purely classical proof[4] that energy is positive for $\Lambda = 0$ gravity can also be applied here. His proof, inspired by the supergravity argument, is based on considering solutions of the Dirac equation $\not{D}\epsilon \equiv \gamma^i D_i \epsilon = 0$ in an external metric satisfying $G_{0\mu} = 0$. From the relations

$$0 = \epsilon^* \not{D}^2 \epsilon \equiv \epsilon^* (D^2 + G_{0\mu}\gamma^0\gamma^\mu)\epsilon = \epsilon^* D^2 \epsilon ,$$

it follows upon integration that

$$\oint dS_i\, \epsilon^* D^i \epsilon = \int |\nabla\epsilon|^2 d^3x \geq 0 . \tag{6.2}$$

The surface integral is then separately shown to be proportional to E, which establishes positivity of the latter. The same reasoning should apply here with D_i

replaced by $\tilde{D}_i$, and the metric now satisfying $G_{0\mu} + \Lambda g_{0\mu} = 0$ provided, as is likely, the surface integral is again proportional to E. Similarly, it would be of interest to generalize the classical geometrical proof of Schoen and Yau[3)] to the $\Lambda < 0$ case. It may even be possible to establish full non-linear stability in the $\Lambda > 0$ case for excitations lying within the horizon by analytic continuation from $\Lambda < 0$, using the static form of the O(4,1) metric which covers the interior region only.

7. CONCLUSIONS

We have seen first that it is possible to parallel all the arguments of $\Lambda = 0$ gravity in a background flat metric in order to establish a satisfactory energy expression in the general $\Lambda \neq 0$ case which is conserved and of flux integral form, as long as the background metric has a timelike symmetry. When applied to the stability problem, these expressions enable one to show that the energy is positive for all asymptotically anti-De Sitter metrics in the $\Lambda < 0$ sector, using methods of supergravity, analogous to the $\Lambda = 0$ case, for grading the algebra. When $\Lambda > 0$, stability of small excitations (about the De Sitter vacuum) which lie within the event horizon was demonstrated. A clear and universal relation between event horizon and Hawking radiation "instability" was then obtained in terms of the general property of any free fields that $T^{00} \geq |T^{0i}|$, together with the simple facts that $|\bar{\xi}_0| < |\bar{\xi}_i|$ beyond the horizon and that $E(\bar{\xi}) = \int (T^{00}\bar{\xi}_0 + T^{0i}\bar{\xi}_i) d^3x$ is the relevant energy. Consequently, contributions to $E(\bar{\xi})$ from beyond the horizon are no longer positive. Semi-classical stability for $\Lambda > 0$ also seems likely, as well as general energy positivity for excitations within the horizon.

We conclude that at least classically, there is no instability argument to rule out the cosmological extensions of Einstein theory, and that they are much like the $\Lambda = 0$ model in this respect.

REFERENCES

1) D. Brill and S. Deser, Ann. Phys. 50 (1968) 548.

2) S. Deser and C. Teitelboim, Phys. Rev. Lett. 39 (1977) 249;
M. Grisaru, Phys. Lett. 73B (1978) 207.

3) P. Schoen and S.T. Yau, Comm. Math. Phys. 65 (1979) 45; Phys. Rev. Lett. 43 (1979) 1457.

4) E. Witten, Comm. Math. Phys. (in press).

5) G.W. Gibbons and S.W. Hawking, Phys. Rev. D10 (1977) 2738.

6) S. Coleman, Phys. Rev. D15 (1977) 2929;
S. Coleman and C.G. Callan, Phys. Rev. D16 (1977) 1762.

7) M.J. Perry in Superspace and Supergravity, eds S.W. Hawking and M. Roček (Cambridge 1981);
D. Gross, M.J. Perry and L. Yaffe, Princeton Univ. preprint (1981).

8) E. Witten, Princeton Univ. preprint (1981).

9) S.W. Hawking and G.F.R. Ellis, The Large Scale Structure of Space-time, (Cambridge, 1973).

10) P.K. Townsend, Phys. Rev. D15 (1977) 2802.

11) S. Deser and B. Zumino, Phys. Rev. Lett. 38 (1977) 1433.

12) R. Arnowitt, S. Deser and C.W. Misner, Phys. Rev. 116 (1959) 1322; 117 (1960) 1595 and in Gravitation: an introduction to current research, ed. L. Witten (Wiley, New York, 1962).

13) H. Nariai and T. Kimura, Progr. Theor. Phys. 28 (1962) 529.

14) D. Brill and S. Deser, Comm. Math. Phys. 32 (1973) 291.

15) G.W. Gibbons and M.J. Perry, Proc. Roy. Soc. A358 (1978) 467.

16) P. Schoen and S.T. Yau, Phys. Rev. Lett. 42 (1979) 547.

17) F. Gürsey and T.D. Lee, Proc. Nat. Acad. Sci. 49 (1963) 179.

WHY IS THE APPARENT COSMOLOGICAL CONSTANT ZERO?

S. W. Hawking
Department of Applied Mathematics
and Theoretical Physics
Cambridge, CB3 9EW
U.K.

ABSTRACT

The apparent cosmological constant is measured to be zero with an accuracy greater than that for any other quantity in Physics. On the other hand one would expect a large induced cosmological constant unless the various contributions from symmetry breaking, etc., were balanced against each other to better than 1 part in 10^{40}. It is suggested that this puzzle can be resolved by assuming that quantum state of the universe is uot chosen at random but contains only states with a very large Euclidean 4-volume. In this situation the actual value of the cosmological constant is unobservable. There are solutions of the Einstein equations with a large cosmological constant which appear nearly flat on large length scales but which are highly curved and topologically complicated on very small length scales. Estimates are made of the spectrum of these topological fluctuations and of their effects on the propagation of particles.

1. INTRODUCTION

Observations of distant galaxies and of the microwave background radiation indicate that the universe is described by a Friedman-Robertson-Walker model to a high degree of accuracy. In such a model the Einstein equations give

$$\frac{3\ddot{R}}{R} = -\frac{4\pi}{m_p^2}\left(\mu + 3p\right) + \Lambda$$

where R is the scale factor, μ and p are the energy density and pressure of the universe, Λ is the cosmological constant and m_p is the Planck mass in units in which $\hbar = c = 1$. In principle the deceleration parameter $q_0 = -\left(\frac{\ddot{R}R}{\dot{R}^2}\right)$ may be determined from the shape of the magnitude-redshift curve for galaxies, but possible evolutionary changes in the brightness of the galaxies mean that we can place only an upper limit on $|q_0|$ of about 2. The universe is dominated by non-relativistic matter at the present time so the effective value of the pressure p is small compared to the energy density μ . Measurements of the density are again a bit uncertain but the matter in the galaxies and clusters of galaxies seems to correspond to a density of about 10^{-30} gm/cc or about 1/10 of the critical density $\frac{3m_p^2\dot{R}^2}{8\pi R^2}$. There might be other forms of matter which have not been observed but it would be very difficult to accommodate more than about ten times the observed amount. Thus one can place an upper limit of about 10^{-32} eV on or about 10^{-60} on $|\Lambda/m_p^2|^{\frac{1}{2}}$.

The effective cosmological constant is thus observed to be zero with an accuracy better than for any other quantity in physics. For example, the observational upper limit on the mass of the photon given by spacecraft measurements of the earth's magnetic field is only about 10^{-16} eV or $m_\gamma/m_e < 10^{-22}$. Even so, we do not believe that the photon mass is so small merely by accident or by the fine tuning of some adjustable parameter. Rather we invoke gauge invariance to make the photon mass exactly zero. By contrast, even if the bare cosmological constant were zero, there does not seem any similar reason why the effective induced cosmological constant should be zero. Indeed one would expect it to be very large for the following reasons:

1) There will be an induced cosmological constant from the diagrams

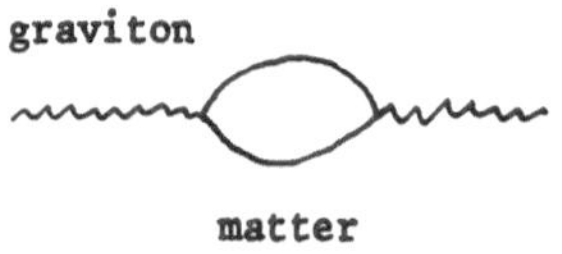

$\sim \frac{\mu^4}{m_p^2}$

where μ is the cut off. The natural cutoff would be the Planck mass. This would give $\Lambda \sim m_p^2$ which would have to be balanced very accurately by a similar negative bare cosmological constant.

2) If there are Higgs scalar fields which break the grand unified or electro weak symmetries, there will be a contribution to Λ of $8\pi V(\phi_c)/m_p^2$ where V is

the effective potential and ϕ_c is the expectation value of the scalar field. It would require very fine tuning of V i.e.

$$\left|\frac{V(\phi_c)}{\phi_c^4}\right| < 10^{-48}$$

in order not to produce a cosmological constant above the observed upper limit.

3) It is believed that at high temperatures QCD behaves like a Coulomb gas but that at about $T \sim 100\ \mathrm{MeV}$ there is a Bose condensation which causes confinement. One might expect that this Bose condensation would give rise to a vacuum energy density of about $(200\ \mathrm{MeV})^4$ or a $\Lambda^{\frac{1}{2}}$ of about 10^{-12}eV. This would have to be cancelled by a bare cosmological term to an accuracy of better than 1 part in 10^{20} in order not to exceed the observed upper limit.

4) In a supersymmetric theory the numbers of bosons and fermions are equal, and their infinite contributions to the cosmological constant cancel each other. However we know from observation that supersymmetry must be broken at energies lower than 10^3GeV. This would give rise to an induced value of $(m_p^2 \Lambda)^{1/4}$ of the same order. One could possibly cancel this positive Λ with a negative Λ arising from gauging the O(N) symmetry in extended supergravity [1] , but this would require very fine tuning.

One way of explaining why the observed cosmological constant is so small would be to appeal to the Anthropic Principle, that the theory which describes the universe must be such as to allow the development of intelligent life: if it did not, there would not be anyone to observe the theory. In the case of the cosmological constant one could argue that a value slightly more negative than the observed limit would have caused the universe to recollapse before life could have developed while a more positive value would have caused the universe to expand too rapidly to allow the formation of galaxies.

Many people, including myself, are rather unhappy about the idea that the theory of physics is especially adjusted to allow for our existence. However most people would agree that, given the physical theory, the quantum state or initial conditions of the universe cannot be entirely arbitrary but must be determined by some requirement. I shall argue that this requirement is that space-time should have some very large, possibly infinite, Euclidean 4-volume, V_o. There exist solutions of the Einstein equations which have arbitrarily large volume, V_o, whatever the actual value of the cosmological constant, Λ [2] . These solutions seem to appear nearly flat on length scales, L, such that $(V_o)^{1/4} \gg L \gg \Lambda^{-1/2}$, but they are curved and topologically complicated on length scales of the order of $(\Lambda)^{-1/2}$. If one did not notice the small scale "foamy" structure, one would think that they were solutions with an apparent cosmological constant of the order of $(V_o)^{-\frac{1}{2}}$. Thus the observed cosmological constant would be zero in the limit that $V_o \to \infty$.

The requirement that V_o be very large or inifinite would seem a natural one to impose on the quantum state of the universe but one could also justify it on anthropic grounds, this time applied to the quantum state rather than to the theory itself: there might be many universes described by quantum states with different values of V_0, but only those for which V_0 was very large would contain intelligent beings to ask why the apparent cosmological constant was so small.

2. THE VOLUME ENSEMBLE

I shall adopt the Euclidean approach [3,4] in which the path integral is evaluated over all positive definite metrics, g, on space-time manifolds of all topologies. If the space-time manifold were not compact, one would have to include a boundary term in the action [5] . We know what this term is for asymptotically flat spaces but it is fairly clear that the universe is not asymptotically flat. I shall therefore consider only compact space-time manifolds. This is not to say that space-time is actually compact but it can be viewed as a normalization condition like periodic boundary conditions. In the class of the spaces with very high topology that I shall be considering, the action of a compact and non-compact manifold will differ only by a relatively small amount.

Let $N(V_0)dV_0$ denote the number of states of the gravitational field with Euclidean 4-volumes between V_0 and $V_0 + dV_0$. One can calculate $N(V_0)$ by inserting a $\delta(V-V_0)$ in the path integral:

$$N(V_0) = \frac{1}{2\pi i}\int_{-i\infty}^{+i\infty} d\left(\frac{m_p^2\Lambda}{8\pi}\right)\int d[g]\, d[\phi]\exp\left\{-I[g,\phi] - \frac{m_p^2\Lambda}{8\pi}(V-V_0)\right\}$$

where $I[g,\phi]$ is the Euclidean action of the positive definite metric, g, and the matter fields, ϕ . The term $\exp\{-\frac{m_p^2\Lambda V}{8\pi}\}$ can be absorbed into the Euclidean action where it acts as a cosmological constant. One can define the volume partition function as

$$Z[\Lambda] = \int d[g]\, d[\phi]\exp\left\{-I[g,\phi,\Lambda]\right\} .$$

Formally, this is the Laplace transform of N(V).

$$Z[\Lambda] = \int_0^\infty N(V)\exp\left\{-\frac{m_p^2\Lambda V}{8\pi}\right\} dV$$

One expects that Log N(V) should be proportional to $V^{\frac{1}{2}}$, so this should converge. The density of states is then given by the inverse Laplace transform.

$$N(V_0) = \frac{1}{2\pi i}\int_{-i\infty}^{i\infty} d\left(\frac{m_p^2\Lambda}{8\pi}\right) Z[\Lambda]\exp\left\{\frac{+m_p^2\Lambda V_0}{8\pi}\right\}$$

The integral in Λ should be taken to the right if the essential singularities in $Z[\Lambda]$ at $\Lambda = 0$.

If there were a bare or induced cosmological constant, Λ_0 , already present in the action, I, it could be absorbed into a shift in the dummy parameter Λ and would give a factor $\exp\{-m_p^2\Lambda_0 V_0/8\pi\}$ in $N(V_0)$. If one calculated physical quantities such as the correlation functions only over states of the gravitational field with a given volume, V_0, one would normalize by this factor. Thus, in this situation, the actual value of the cosmological constant would be unobservable.

One might think that the volume of a solution of the Einstein equations would be of order $|\Lambda|^{-2}$. However, the examples of the Einstein-Kahler metrics [2] show that there can be solutions with very large volumes and very high Euler numbers, χ. These spaces seem to appear nearly flat on length scales, L, such that $(V_0)^{1/4} \gg L \gg (V_0/\chi)^{1/4} \sim \Lambda^{-1/2}$. Their action is of the form

$$I[g,\Lambda] = -\frac{m_p^2 a}{8\pi\Lambda} + b$$

where 'a' and 'b' depend on the solution but not on Λ (at the tree level). The "a" term arises from the usual Einstein Lagrangian $a \geq 0$ $a \sim c\chi$ for large χ, where $c \sim O(1)$. The "b" term is non-zero if there are quadratic curvature terms in the action, $b \geq 0$ for stability, and $b \propto \chi$ for large χ.

One can show how these solutions contribute to $Z[\Lambda]$ by considering the Borel transform [6].

$$Z[\Lambda] = \int_0^\infty B(x) \exp\left\{-\frac{m_p^2 x}{8\pi\Lambda}\right\} d\left\{\frac{m_p^2 x}{8\pi\Lambda}\right\}$$

The Borel transform B(x) will have singularities at the values of x equal to the quantity (-a) in the classical solutions. Thus $Z[\Lambda]$ is Borel summable because all the singularities of B(x) lie on the negative x-axis apart from a possible δ-function contribution to B from the singularity at $x=0$ corresponding to the K-3 solution.

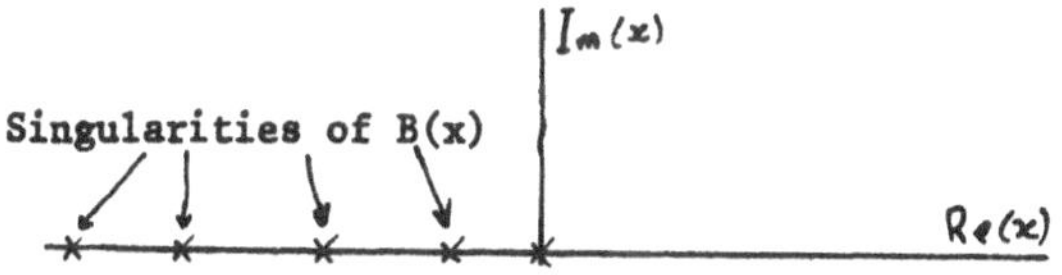

If B(x) is suitably behaved for large x, this implies that $Z[\Lambda]$ is non-singular for Re $\Lambda > 0$, and that N(V) = for V < 0. One can express N(V) in terms of B(x) by

$$N(V) = (m_p^2/8\pi)^2 \int_0^\infty B(x)\, J_0\left(\frac{m_p^2 x^{1/2} V_0^{1/2}}{4\pi}\right) dx$$

3. SUPERGRAVITY

The action for N=1 supergravity can be written as an integral of the curvature superfield R over chiral superspace [7].

$$I = \frac{m_p^2}{8\pi} \int d^4x\, d^2\theta\, R \det \tilde{E}$$

where E is the restriction of the achtbein to the chiral subspace. One can add a term

$$(m_p^3 e/8\pi)\int d^4x\,d^2\theta\,\det\tilde{E}\ ,$$

proportional to the volume of chiral superspace, where e is a dimensionless coupling constant. This gives a theory with a cosmological constant $\Lambda = -e^2 m_p^2$.

One can now consider the number of states $N(V_0)\,dV_0$ with chiral supervolumes between V_0 and $V_0 + dV_0$.

$$\mathcal{N}(V_0) = \frac{1}{2\pi i}\int_{-i\infty}^{i\infty} d\left(\frac{m_p^3 e}{8\pi}\right) \exp\left\{\frac{m_p^3 e\, V_0}{8\pi}\right\} Z[e]$$

where

$$Z[e] = \int d[\mathcal{U}] \exp\left\{-I[\mathcal{U}, e]\right\}.$$

In a similar manner one can define the Borel transform by

$$Z[e] = \int_0^\infty B(x) \exp\left\{x/(8\pi e^2)\right\} d\left(x/(8\pi e^2)\right).$$

The argument goes through much as in the ordinary gravity case. There is however a difference in that one might expect the fermionic fluctuations to cancel the bosonic ones and so make $Z[0] = 1$. In this case

$$1 = \int_0^\infty \mathcal{N}(V)\,dV$$

This would suggest that $N(V) \to 0$ as $V \to \infty$. On dimensional grounds one might guess

$$\mathrm{Log}\left\{\mathcal{N}(V)\right\} \propto -m_p^2 V^{2/3}.$$

We do not yet have a superspace formulation of the higher N supergravities, but if the N=8 theory can be derived by dimensional reduction from N=1 supergravity in 11-dimensions, the volume of majorana superspace

$$\int d^{11}x\,d^{16}\theta\,\det\tilde{E}$$

has the right dimensions of $[x]^3$ to be added to the action as a cosmological term. In fact all that one needs is that it should be possible to write the action for the gauged O(N) theory as

$$I[e] = I[0] + (m_p^3 e/8\pi)\,J$$

where J is some invariant integral over superspace. One can then consider $\mathcal{N}(J_0)\,dJ_0$ the number of states between J_0 and $J_0 + dJ$.

4. THE SMALL-SCALE STRUCTURE OF SPACETIME

In order to define what it means to say that space-time has a certain structure on a length scale, L, suppose (for the moment) that space-time is a smooth compact manifold with positive definite metric. Then one can define a distance function, $d(x,y)$, and can cover the manifold by a finite collection, $C(L)$, of balls of radius L. One can regard C(L) as a simplicial complex with a topology given by Cěch Cohomology. Then one can define the Euler number, $\chi(L)$, as the minimum χ of these complexes. One is interested in $\chi(L)/V$, the density of Euler number on the scale L.

A. Higher Derivative Theories

If the action contains terms quadratic in the curvature, these will dominate over the Einstein and cosmological terms at short distances. If the coefficients of the higher derivative terms were constant, the Euler number density would be scale invariant, i.e. proportional to L^{-4}. However it seems that one loop effects will cause these coefficients to increase at small L [8]. This will damp out topological fluctuations below some scale, L_0, and the space-time manifold will be smooth below this scale.

B, Supergravity

If the effective action is just the classical action, one would expect the Euler number density to go up two powers of L^{-1} faster than in the scale invariant case, i.e. to be of order L^{-6}. In this case space-time would be a fractal and not a smooth manifold. However there could be terms of order $(\text{curvature})^4$ and higher in the effective action even if the theory is finite to all orders. These might provide a cut-off to topological fluctuations below some length-scale, L_0.

One can ask how particles would propagate through such a foamy space-time. On general grounds and from some particular examples it seems as if a topological fluctuation of scale, ρ, will scatter a particle of spin, , and momentum, $k \ll m_p$ by an amount [9]

$$\rho^2(\rho k)^{2s}$$

topological fluctuation.

If the spectrum of topological fluctuations is $\sim (m_p L)^{-\gamma} m_p^4$ for $m_p^{-1} > L > L_0$, then scalar particles will acquire an effective mass of order $[(m_p L_0)^{-\gamma} - 1] m_p$ but fermions will be protected by gauge invariance from acquiring a mass.

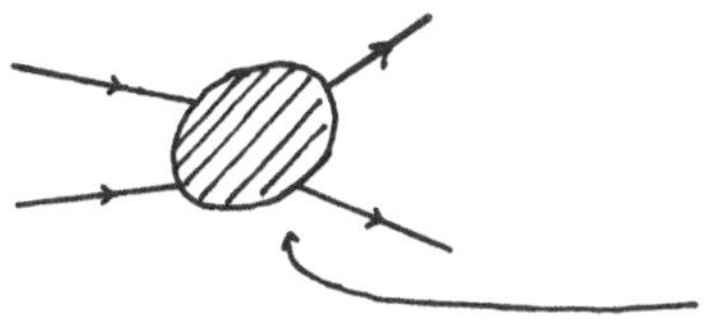

topological fluctuation.

The simultaneous scattering of two particles by a topological fluctuation will give rise to an effective 4-point vertex of the order of

$$m_p^{-4s}\left[(m_p L_0)^{4+4s-\gamma} - 1\right]$$

If $\gamma = 4$, there will be very little effect on particles of spin $s \geq \frac{1}{2}$, even if $L_0 = 0$, i.e. even if there is no cut-off to the fluctuations. If $\gamma = 6$, the vertex will be small for $s \geq 1$ but will be logarithmically divergent as $L_0 \to 0$ for $s = \frac{1}{2}$

It follows from these results that if there is a lower limit, L_0, the scale of topological fluctuations, then scalar particles will acquire a very large mass, but particles of spin, $s \geq \frac{1}{2}$, will propagate almost as if in flat space. Any observed scalar particles would have to be bound states. The same conclusions would hold even if there were no lower cut-off, L_0, to the scale of topological fluctuations, provided that $\gamma < 6$. However, if $\gamma = 6$, then the observed spin-½ particles would also have to be bound states. In this case, one might believe, like Ellis, Gailland and Zumino [10] that all observed particles, except possibly the graviton, were bound states or collective excitations of the foam. Such excitations would form representations of the Poincaré group corresponding to the almost flat structure of space-time on scales larger than the Planck length. They would thus appear as particles of spin, S = 0, ½ and 1. Higher spin excitations would not obey re-normalizable effective field theories and so might not be observable at low energies.

Acknowledgements

I would like to thank Murray Gell-Man for arousing my interest in the problem of the cosmological constant and Bernard Whiting for help in preparing and delivering this talk.

REFERENCES

1. S. Ferrara, E. Cremmer, B. Julia, J. Scherk, P. van Nieuwenhuizen & L. Girardello. Nuc.Phys. B147, 105, (1979).

2. S. W. Hawking, Nuc. Phys. B144, 349 (1978).

3. S. W. Hawking, in General Relativity : An Einstein Centenary Survey ed. S. W. Hawking & W. Israel, Cambridge University Press, Cambridge. 1979.

4. S. W. Hawking, in Recent Developments in Gravitation, Plenum Press (1979).

5. G. W. Gibbons & S. W. Hawking, Phys.Rev, D15, 2752-6 (1977).

6. J. Zim-Justin, Phys.Reps. 70, 109-167 (1981).

7. W. Siegel & J. Gates, Nuc.Phys. B147, 77-104 (1979).

8. E. S. Fradkin & A. A. Tseytlin, Higher Derivative Quantum Gravity: One-Loop Counterterms & Asymptitic Freedom, Lebedev Institute preprint 1981.

9. S. W. Hawking, D. N. Page & C. N. Pope, Nuc.Phys. B170, 283-306 (1980).

10. J. Ellis, M. K. Gaillard & B. Zumino, Phys.Letts. 94B, 343-348 (1980).

LATTICE GRAVITY

OR

RIEMANNIAN STRUCTURE ON PIECEWISE LINEAR SPACES[1)]

J.Cheeger[2)5)]

W.Müller[3)]

R.Schrader[4)6)]

Institut des Hautes Etudes Scientifiques

91440 Bûres-sur-Yvette, France

1) Extended version of a talk presented by R.S.at the Heisenberg Symposium, München 1981.

2) Permanent address: Math.Dept.SUNY at Stony Brook, N.Y.11794 U.S.A.

3) Permanent address: Zentralinstitut für Mathematik und Mechanik, Akademie der Wissenschaften der DDR, Berlin, DDR.

4) Permanent address: Institut für theoretische Physik, Freie Universität Berlin, Berlin (West).

5) Supported in part by the National Science Foundation, NSF Grant MCS 7802679 A02.

6) Supported in part by the Deutsche Forschungsgemeinschaft.

Lattice formulations of various quantum field theoretic models including gauge theories have turned out to be useful high energy cut-off versions (see e.g. the review talk of K.Symanzik in [1]). They permit the application of techniques, like analytical, nonperturbative methods of statistical mechanics, as well as numerical Monte Carlo Calculations.

Here we wish to show that Regge calculus ([2]) provides the possibility of discussing gravity in a similar spirit. This has also been realized independently by J.Fröhlich ([3]).

The actual mathematical results we wish to announce can be viewed as classical approximation theorems for the Regge calculus (details will appear elsewhere). Since such theorems typically pick out appropriate lattice Lagrangeans, they constitute the intial step in our program. If successful, the approach would eventually lead to a quantization of gravity within the context of euclidean path integral formulation, the attractive features of which are well known. It combines Lagrangean quantum field theory, statistical mechanics and axiomatic quantum field theory (see e.g.[4] ,[5] [6]). Therefore, as has been stressed by Gibbons and Hawking ([7]), (see also the review article by Hawking in [8]), it is tempting to try to take advantage of this approach for the case of gravity.

In a euclidean path integral formulation of pure gravity, the basic objects of the theory are riemannian spaces (M,g), where g denotes the metric of the manifold M. Invariants, such as the volume $(M,g) \longmapsto \mathrm{Vol}(M,g)$, the total scalar curvature $(M,g) \longmapsto R(M,g)$, lengths of curves etc., can then be viewed as random variables, i.e. functions on a probability space. For fixed M(henceforth assumed to be compact), the associated probability space should be obtained from the moduli space $\mathcal{S}=\mathcal{M}/G$, where $\mathcal{M}$ is the set of all metrics on M and G is the diffeomorphism group of M (see e.g. the articles of D.G.Ebin and H.Omori in [9] for an analysis of their structures). An open dense subset of $\mathcal{S}$ may be given the structure of an infinite dimensional riemannian space. If an extrapolation from the finite dimensional case were allowed, this would give rise to a volume form which, when combined with a Gibbs factor, for example of the form $\exp(-\lambda_1 R(M,g) - \lambda_2 \mathrm{Vol}(M,g))$ (λ_1, λ_2 bare coupling constants), would produce the desired measure. Note that the stationary points of the Gibbs factor give classical solutions, i.e. solutions of Einstein's equations (with a cosmological term). Actually, we expect this measure to live on something like $\mathcal{S}' = \mathcal{M}'/G$ where $\mathcal{M}'$ is the set of"distributional metrics" on M. Also, to allow for local fluctuations in topology (the "foam structure" of Wheeler [10]),one eventually wants to let the topology of M vary as well.

Now up to this point, the discussion has been formal, a familiar phenomenon when one deals with an infinite number of degrees of freedom. This suggests focusing attention on riemannian spaces which are described by only a finite number of parameters. Of necessity, such spaces will only be smooth almost everywhere. The simplest spaces of this type are piecewise linear (or, one might say, piecewise flat) spaces. The topological properties of these spaces have been intensively studied by mathematicians for many years. Their geometric and analytic properties have also been studied (see e.g. [11] [12] [13] [14] [15] [16] [17] [18] [19] [33] [34] [35]) but up till now, have not received equal attention.

To describe piecewise linear spaces, we recall some elementary definitions (see e.g. [20] for further details). An i-simplex, σ^i, is the convex hull of a set of (i+1) points, $v_0 \ldots v_i$, of $\mathbb{R}^N$, which lie in no (i-1)-dimensional affine subspace. The v_j are called vertices of σ^i and $\sigma^i = \sum_{j=0}^{i} \alpha_j v_j$ where $\alpha_j \geq 0$, $\Sigma\, \alpha_j = 1$.

Simplices, whose vertices are subsets of $v_0, \ldots, v_i$, are called faces of σ^i. One dimensional faces σ^1 are called edges.

A simplicial complex K is the space formed by a (finite) collection of simplexes in $\mathbb{R}^N$ with the property that any two intersect in a common face (or are disjoint). More abstractly we can view a simplicial complex as a space formed by "gluing together" a collection of disjoint simplices by linear isomorphisms of certain faces. If $K \subset \mathbb{R}^N$ as above, or if the gluing maps are distance preserving, then K inherits a natural metric structure (i.e. it is a metric space). For a fixed pattern of identifications (combinatorial structure), any such metric structure is determined by specifying the finitely many edge lengths $\ell(\sigma^1)$, subject to the appropriate strict (triangle) inequalities. By analogy with (M,g) we write (K,ℓ) to indicate the metric structure on $K(\ \ell = \{\ell(\sigma^1)\}_{\sigma^1})$.

The union of all i-simplices of K is called the i-skeleton, Σ^i. The i-skeleton of (K,ℓ) has a natural volume $V^i_{(K,\ell)} = \sum_{\sigma^i \in \Sigma^i} V(\sigma^i)$ where $V(\sigma^i)$ is the i-dimensional volume of σ^i.

If moreover, K is topologically equivalent to an n-dimensional manifold with empty boundary then its deviation from flatness (curvature) can be measured by assigning a defect $\delta(\sigma^{n-2})$ to each (n-2) simplex σ^{n-2}. K is locally isometric to Euclidean space if and only if all defects vanish. To define $\delta(\sigma^{n-2})$ we first introduce the notation $[\sigma^i, \sigma^j]$ for the dihedral angle at the face σ^i of σ^j. Here we mean the angular measure of the unit normals to σ^i at $p \in \sigma^i$, which point into σ^j. We use the normalization for which all unit spheres have measure 1.

We then set

$$\delta(\sigma^{n-2}) = 1 - \Big(\sum_{\sigma^n \supset \sigma^{n-2}} [\sigma^{n-2}, \sigma^n] \Big) \qquad (1)$$

This defect may also be defined in a more intrinsic manner as follows. We start at $p \in \sigma^{n-2}$ and walk a distance ε(ε small) directly away from σ^{n-2}. We then walk in a circle of length $2\pi\beta\varepsilon$ staying at constant distance from σ^{n-2}. Then $\delta(\sigma^{n-2}) = 1-\beta$. To understand the defect intuitively , consider a 2-dimensional example M^2 such as the surface of a cube.

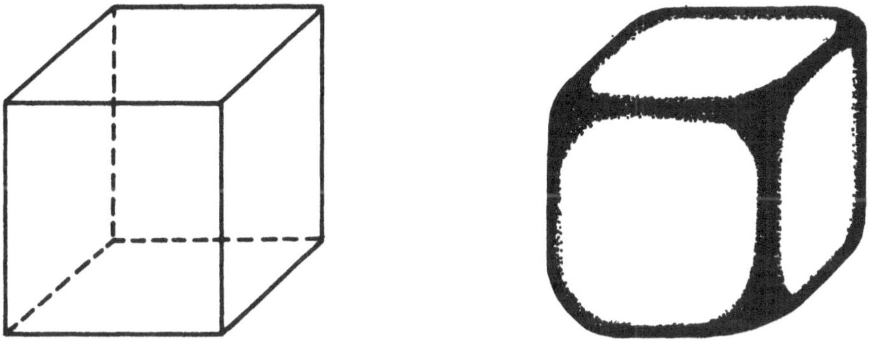

Fig.1

The defect at each vertex v_i is $\delta(v_i) = \frac{1}{2\pi}(2\pi - \frac{3}{2}\pi) = \frac{1}{4}$. By "rounding the corners and edges with sand paper" one sees that the surface of the cube can be viewed as having curvature in form of a Dirac δ-function of strength $2\pi\ \delta(v_i) = \frac{\pi}{2}$ concentrated at each corner, and thus a total curvature of 4π. Since the surface of the cube is homeomorphic to that of a sphere, this coincides with the total curvature predicted by the Gauss-Bonnet theorem, provided we define the total curvature at a vertex as $2\pi\ \delta(v_i)$.

For higher dimensions T.Regge([2]) proposed the expression

$$R^c_{(K,\ell)} = \sum_{\sigma^{n-2} \in (K^n,\ell)} \delta(\sigma^{n-2}) v(\sigma^{n-2}) \qquad (2)$$

as the analogue of the total <u>scalar</u> curvature. More generally one can assign to each open set U the sum over those σ^{n-2} in U and obtain the analogue of the scalar curvature density Rdvol. Note that if all lengths in (K^n,ℓ) are multiplied by a constant k , angles do not change. Thus the Regge scalar density gets multiplied by k^{n-2}, the scaling property which also holds for the ordinary scalar curvature density. To gain further confidence in Regge's definition, one would like to be able to claim that if (K^n,ℓ) approximates a smooth riemannian manifold M^n in a suitable sense, then R^c approximates R. This assertion, for which Regge gave only a heuristic argument, seems to have long been assumed correct by those with an interest in the subject ([21]).We will now examine it in some detail. To form a p.l. approximation to M^n, one can start with a smooth triangulation, i.e. a piecewise smooth topological equivalence between M^n and some simplicial complex K. Such triangulations exist and are unique up to a subdivision (see e.g. [22]). Intuitively triangulating M^n simply means cutting it up into disjoint curvilinear simplices. Fig.2a is an example for the twodimensional sphere.

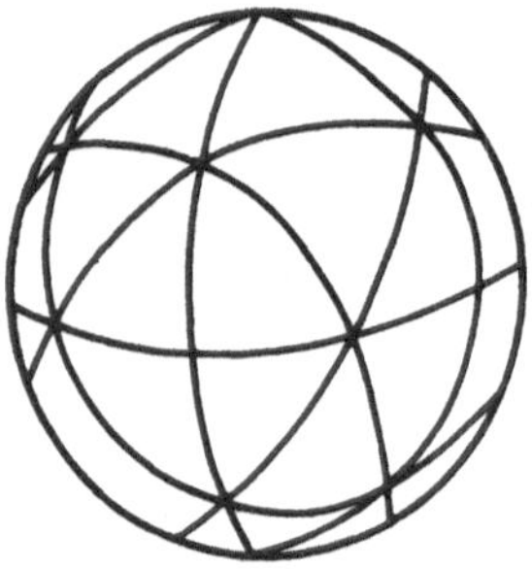

Fig.2a

Fig.2b

If M^n is riemannian,it is not difficult to see that the edges can be chosen to be geodesics. If the mesh η (the maximum edge length) is sufficiently small and if the simplices are sufficiently "fat", one sees that there exists a piecewise linear simplicial complex (the intrinsic approximation) with the same combinatorial structure and edge lengths (the fatness of a simplex is the ratio of volume to the n-th power of the longest edge length). Alternatively, if M^n is given as a subset of some $\mathbb{R}^N$, one can form the secant (or extrinsic) approximation. This is the simplicial complex consisting of the linear simplices in $\mathbb{R}^N$ with the same vertices as those of the curvilinear simplices of M^n (figure 2b is an example for the two-dimensional sphere). Although it is not completely obvious that the two methods are equivalent for the problem at hand,using Regge's formula (6) in [2], one sees that this is indeed the case. Let M^n_η denote a sequence of intrinsic approximations of mesh $\eta \to 0$ and fatness $\theta \geqslant \theta_0 > 0$. In dimensions 2 the convergence

$R^c_\eta = R + O(\eta)$ is a relatively simple consequence of the Gauss-Bonnet formula for the excess $e(\sigma^2)$ of a geodesic triangle $\sigma^2 \subset M^2$. If α_j are the angles of σ^2 and $\bar{\alpha}_j$ the corresponding angles of the flat simplex $\bar{\sigma}^2$ with the same edge lengths, we easily obtain

$$\sum_{v_j} R^c(v_j) = \sum_{\sigma^2} e(\sigma^2) + O(\eta) \, , \tag{3}$$

$$\int_{\sigma^2} K \, dA = e(\sigma^2) \overset{\text{def}}{=} \sum_{j=1}^{3} (\alpha_j - \bar{\alpha}_j) \, . \tag{4}$$

If this argument were to generalize to higher dimensions, one would expect a definition of $e(\sigma^n)$ for which

$$\sum_{\sigma^{n-2}_\eta} R^c(\sigma^{n-2}_\eta) = \sum_{\sigma^n_\eta} e(\sigma^n_\eta) + O(\eta) \, , \tag{5}$$

and at least an equation like

$$\int_{\sigma^n_\eta} R \, dvol = e(\sigma^n_\eta) + o(\eta^n) \, . \tag{6}$$

In fact, a definition of $e(\sigma^n)$ which gives (5) is not hard to come by. However, when one calculates $\lim_{\eta \to 0} e(\sigma^n_\eta) \cdot \mathrm{Vol}(\sigma^n_\eta)^{-1}$ one obtains an expression, which, while linear in the curvature, is definitely not equal to the scalar curvature R evaluated at some point of σ^n_η. The limit depends on the (limiting) proportions and spatial orientation of σ^n_η. In spite of the failure of the above local calculation, the more global equation

$$\lim_{\eta \to 0} \sum_{\sigma^{n-2}_\eta \subset U} R^c(\sigma^{n-2}_\eta) = \int_U R \, dvol + O(\eta) \tag{7}$$

turns out to be correct. In other words R^c_η converges to R as a measure.

We have two methods of proof. Each works directly with the extrinsic approximation and each involves showing that R^c plays a specific role for p.l. spaces analogous to that played by scalar curvature in the smooth case. Our first argument depends on the generalization to p.l. spaces of the kinematic formula of Blaschke, Santaló and Chern in integral geometry (see e.g. [23][24]). The second depends on a generalization of Weyl's formula for the volume of tubes [25]. In both cases, we were stimulated by the work of P.Wintgen ([26]). The kinematic formula and tube formula actually involve a whole sequence of total curvatures, the so called Lipschitz Killing curvatures.

These include the scalar curvature and, in even dimensions, the Gauss-Bonnet form (more correctly "density") as extreme cases. The formulas also apply to manifolds with boundary in which case in particular the mean curvature arises (for the mean curvature, one replaces 1 by $\frac{1}{2}$ in rel.(1)). Quite naturally there are classical approximation theorems like rel.(7) for these higher order curvatures as well. As an example, we write the Gauss-Bonnet density for even dimensional p.l. manifolds (there is a simple generalization to arbitrary simplicial complexes). We have

$$\chi(K) = \sum_{\sigma^0 \subset \sigma^{2\ell_1} \subset \ldots \subset \sigma^{2\ell_j}} (-1)^j \left[\sigma^0, \sigma^{2\ell_1}\right]\left[\sigma^{2\ell_1}, \sigma^{2\ell_2}\right] \ldots \left[\sigma^{2\ell_{j-1}}, {}^{2\ell_j}\right] \tag{8}$$

Here the summation is over all ascending chains of even dimensional simplices. A more elementary formula for $\chi(K)$ in terms of exterior dihedral angles is given in [12]. Although it is possible to derive rel.(8) directly from this last formula, it is interesting to note that the original derivation of rel.(8) (and of the analogous formulas for the other Lipschitz Killing curvatures) was by means of study of the asymptotic expansion of the trace of the heat kernel on p.l. spaces (see [18] [19] [27]). In both the smooth and p.l. cases, it turns out that the Lipschitz Killing curvatures are given by (the same) specific linear combinations of the coefficients on i-forms. When this interpretation is combined with the results announced here, one obtains approximation theorems for these <u>specific</u> combinations of coefficients. (For other combinations, such approximation definitely does not take place). We also mention that a formula for Pontryjagin classes also arises in the context of the heat kernel, but it is of a quite different character from rel.(8), see [19]. Especially in the context of relativity (gravity), it is natural to try to extend the above discussion to Lorentz manifolds. We are presently investigating this problem.

The preceeding discussion should serve to emphasize that p.l. spaces are <u>not</u> themselves discrete objects and, in that respect, differ from usual lattice type approximations. They are in fact, continuous geometric objects, of interest in their own right, which happen to be describable by a finite number of parameters.

On the other hand, one can introduce lattice scalar fields, Higgs fields and Yang Mills fields on p.l. spaces in a spirit closer to standard lattice approximations ([28]) and allow for quantities like Wegner-Wilson loops etc.([29], [28]). Again there are corresponding approximation theorems for Lagrangeans like the p.l. version of the Wilson lattice action ([28]), by which now Yang-Mills fields not only are selfcoupled (non-abelian case) but also interact with gravity. We also expect that spinor fields can be defined on suitable p.l. spaces (see also [3]), and it would be interesting to see whether supersymmetric structures could be introduced. We outline the possibility of doing statistical mechanics of p.l. spaces (we restrict ourselves to the case of pure gravity, the inclusion of other fields being conceptually analogous). For fixed K, we start with partition functions like

$$Z\,(K,\lambda_1,\lambda_2) = \int_{Tr(K)} \exp(-\lambda_1 R^c\,(K,\ell) - \lambda_2 V\,(K,\ell))\,d\ell \qquad (9)$$

Here $Tr(K) \subset \mathbb{R}^L$ (L = # of edges) is the nonempty, open set of ℓ's such that (K,ℓ) is a p.l. space and $V(K,\ell) = V^n(K,\ell)$ ($n = \dim K$).

Also

$$d\ell = F(\ell) \prod_{\sigma^1} d\ell^2(\sigma^1) \qquad (10)$$

should be the discrete version of the formal volume form for S mentioned in the introductory remarks.

So the first problem is to find $F(\ell)$. This is achieved in the following way. Let $v_1 \dots v_n$ be a basis of unit vectors of $\mathbb{R}^n$ forming the edges of an equilateral simplex ($\langle v_i,v_i\rangle = 1$, $\langle v_i,v_j\rangle = \frac{1}{2}$ $(i \neq j)$). Let $\hat{\mathcal{S}}$ be the space of ordered simplices with given edge lengths and $\sigma^n \varepsilon \hat{\mathcal{S}}$. Then the matrix

$$g_{ij} = \frac{1}{2}\,(\ell^2(o,i) + \ell^2(o,j) - \ell^2(i,j)) \quad (\ell(i,i)=0) \qquad (11)$$

of inner products of edges emmanating from the smallest vertex 0 of σ^n (the others being numbered 1,...n) is positive definite. Define an isomorphism from the isometry classes of such σ^n into the positive inner products $\mathcal{P}$ on $\mathbb{R}^n$ by $\sigma \to B_\sigma$ where $B_\sigma(v_i,v_j) = g_{ij}$. $\mathcal{P}$ is a riemannian manifold in a natural way (it is a symmetric space). The pull back of its metric to $\hat{\mathcal{S}}$ via the above isomorphism gives a metric which is independent of the ordering of the vertices. Thus it induces a metric on the space $\mathcal{S}$ of unordered n-simplices. Now fix the combinatorial structure on K and let the metric structure (K,ℓ) vary. Then there is a natural imbedding $(K,\ell) \to \prod_{\sigma^n \varepsilon K} \mathcal{S}$. The riemannian metric on the collection of all (K,ℓ) is induced from this imbedding. Since we are in a finite dimensional situation, there is an associated volume element. As an example, if K is an n-simplex σ^n, then it is easy to see that $F(\ell) = \mathrm{Vol}(\sigma^n)^{-(n+1)}$. The next question that arises is to find the situations where $Z < \infty$. In familiar contexts $Z < \infty$ is a consequence of dynamical stability (the semiboundedness of the Hamiltonian) giving a vacuum state, i.e. a ground state or zero temperature state of the theory. Due to the form of the function $F(\ell)$, (in general $F(\ell)^{-2}$ is a polynomial in the $\ell^2(\sigma^1)$ of degree 2L such that $d\ell$ is scale invariant), the integral in (9) will be divergent for small ℓ's, i.e. short distance singularities still show up in this lattice formulation. Therefore, at this stage, it is necessary to introduce a short distance cut-off $\kappa > 0$, i.e. we only consider those ℓ's in $Tr(K)$ with $Vol(\sigma^k) \geq \kappa^k$ for all $\sigma^k \in K$ ($1 \leq k \leq \dim K$). Call this set $Tr_\kappa(K)$ and denote by Z_κ the corresponding partition function. In the examples (see below) we have examined so far, it turns out that the Gibbs factor acts as a damping factor for large ℓ's , so the problem $Z_\kappa < \infty$ (pure gravity) has the equivalent formulation:

For fixed K,κ and suitable λ_1,λ_2 are $R^c(K,\ell)$ or $V(K,\ell)$ (or other curvature invariants) sufficiently strong infrared regulators? As for the first example, let K be a triangulation of $\mathbb{R}^2$ as depicted in figure 3, O any bounded domain and $K^O = K \cap O$ the full subcomplex of K contained in O :

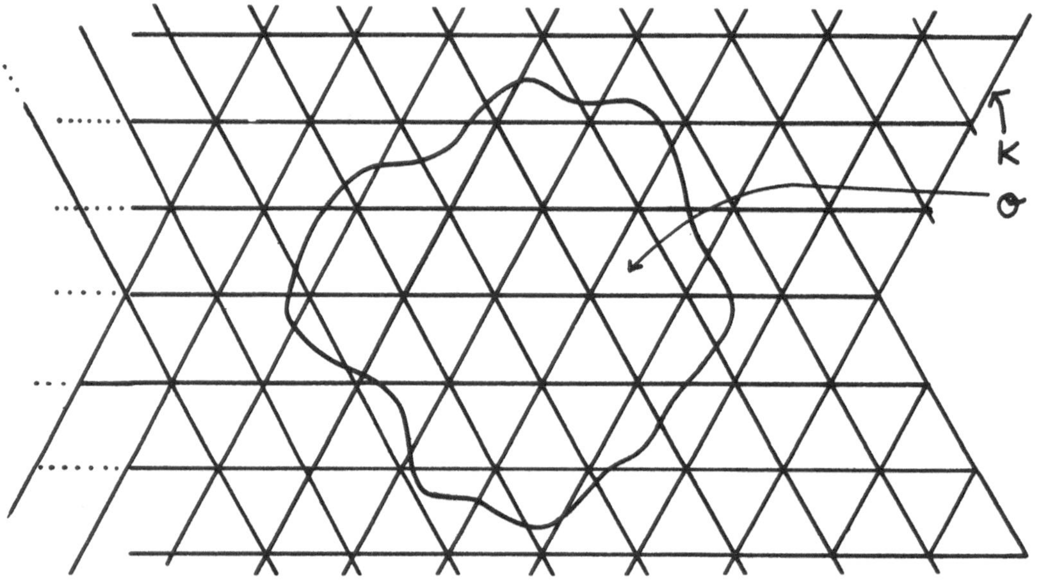

Fig.3

Then $Z_\kappa(K^O,\lambda_1,\lambda_2) < \infty$ for $\lambda_2 > 0$. Also $Z_\kappa(K,\lambda_1,\lambda_2) < \infty$ ($\lambda_2 > 0$) if K is the boundary of the standard 3 simplex. Conversely $Z_\kappa(K,\lambda_1=0,\lambda) = \infty$ for all λ_2 if K is the standard 3-simplex. For the case $K = \partial$ (standard 4-simplex) such that L= # of edges = 10, numerical evidence from computer calculations carried out with A. Karpf and E. Tränkle [30] indicates that $Z_\kappa(K,\lambda_1,\lambda_2) < \infty$ whenever $\lambda_1 \geq 0, \lambda_2 \geq 0, \lambda_1+\lambda_2 > 0$.

Once it is known that the partition function is finite, one has a probability theory with $Tr_\kappa(K)$ as probability space and

$$Z_\kappa^{-1} \exp(-\lambda_1 R^c(K,\ell) - \lambda_2 V(K,\ell)) \mathcal{D}\ell \tag{12}$$

as probability measure. Then $V(K,\ell)$ and Lipschitz-Killing curvatures can be viewed as random variables. Lengths of curves can also be made into (local) random variables as follows: For fixed K, choose the p.l. space $\underline{K} = (K,\ell(\sigma^1)=1)$ as a reference space and let γ be a piecewise differentiable curve in $\underline{K}$. For any choice of metric structure (K,ℓ) there is an obvious (identity) map T_ℓ from $\underline{K}$ to (K,ℓ). Then $\gamma(\ell)$ = length of $T_\ell(\gamma)$ becomes a random variable. In the same way other random variables may be constructed.

This probabilistic formulation immediately leads to two problems. The first is the existence problem of the thermodynamic (bulk) limit. In the example shown in Fig.3 above, it corresponds to asking what happens to expectation values of local observables when O tends to infinity in the sense of van Hove (see e.g. [11]). This leads to the question: Are there situations where phase transitions occur and if so what do they mean? The existence of critical points and their structure is important for the second main problem which goes under the name scale limit. It shows up when one a) considers a sequence of subdivisions of a given simplicial complex K and when one b) tries to take the limit $\kappa \to 0$. This leads to renormalization problems and the number of subdivisions may serve as a renormalization group parameter. Of course, by standard criteria, pure gravity is a nonrenormalizeable theory (for a recent discussion, see e.g. the article of S.Weinberg in [8]). But the present setup allows one to reconsider the problem in the compact case, since previous discussions were within perturbation theory with asymptotic flat metrics, the obvious situation for scattering theory of gravitons.

The scale limit, if it exists, should exhibit what Wheeler has called the foam structure ([10]): By considering finer and finer subdivisions, one allows local fluctuations of curvature and topology on smaller and smaller scales. Note that at the boundary of $Tr(K) \subset \mathbb{R}^L$ the topology of (K,ℓ) changes. Of course this set is of Lebesgue measure zero but it is a priori not clear what happens in the scale limit. There is another effect which we expect to be related to the scale limit: Note that the triangulation of a given riemannian manifold acts like a coordinatization in the sense that one breaks the coordinate gauge invariance. Experience from known quantum field theoretic models, however, leads us to expect that this invariance should reappear in the scale limit and manifest itself through the fact that the limit is independent of the particular sequence of (fat) subdivisions chosen.

Finally we would like to mention that Lagrangians such as $R^c(K,\ell)$ and $V(K,\ell)$ have local properties known in other contexts to lead to Markov property and physical positivity.

At first sight our discussion might lead to the impression that we have been able to find (within this lattice context) a local field theory involving (coordinate) gauge independent quantities only, namely the geodesic lenghts. However, there is the factor $F(\ell)$ in rel.(10) which prevents this. By construction, $F(\ell)$ is the square root of a determinant of a matrix with local building blocks, but $F(\ell)$ itself (and $\log F(\ell)$) is a highly nonlocal object.

Using the standard trick of introducing Faddeev-Popov ghosts, $F(\ell)$ can be represented by a local Lagrangien in the Gibbs factor. In other words, locality is restored at the cost of introducing additional degrees of freedom. We consider this a satisfactory aspect of our approach since it shows that all known and important difficulties with quantum gravity are also present in this formulation. It would be interesting to see whether Markov properties, first discovered by Nelson in the context of euclidean quantum field theories (see his contribution in [4]) hold in this formulation. A related problem is the question of how to do Monte Carlo calculations using e.g. heat bath methods. The presence of fermions, namely the Faddeev-Popov ghosts, of course create well known difficulties.

On the other hand, the introduction of ghost fields allows one to prove the physical positivity, which in contexts discussed so far gives the Hilbert Space of the corresponding relativistic quantum field theory [32] . Indeed, it will hold whenever there is a reflection structure on K (see also [3]). By this we mean a pair (K_+, ϑ) where K_+ is a full subcomplex of K and ϑ a simplicial involution of K such that on $K_+ \cap \vartheta K_+$, ϑ is the identity. Intuitively, K_+ , ϑK_+ and $K_+ \cap \vartheta K_+$ represent the future, the past and the present respectively, while ϑ is the time reflection. In this context, it might be tempting to speculate whether a quantum theory of gravity necessarily requires time reflection to be a symmetry operation (see e.g. the article by Penrose in [8]).

Acknowledgement: It is a pleasure to thank Prof.Kuiper for his hospitality at the I.H.E.S. which made our collaboration possible. Also one of us (R.S.) would like to thank Prof.Yang for his hospitality at the Institute for Theoretical Physics at Stony Brook. We are indebted to J.Fröhlich, M.Gromov, H.Leutwyler, J.Milnor, P.Trauber and P.Wintgen (†) for moral and technical support.

REFERENCES

[1] R.Schrader, R.Seiler, D.A.Uhlenbrock, eds.,Proceedings of the 1981 IAMP Conference in Berlin, Lecture Notes in Physics, Springer, Berlin-Heidelberg-New York, to appear.

[2] T.Regge, Nuovo Cimento 19,551-571 (1961).

[3] J.Fröhlich, Regge calculus and discretized gravitational functional integrals, I.H.E.S. preprint, to appear.

[4] G.Velo and A.Wightman, eds.,Constructive Quantum Field Theory, Erice Summer School 1973, Lecture Notes in Physics, Vol.25, Springer, Berlin-Heidelberg-New York, 1973.

[5] B.Simon, The $P(\phi)_2$ Euclidean (Quantum) Field Theory, Princeton University Press, Princeton, 1973.

[6] J.Glimm and A.Jaffe, Quantum Physics,Springer, Berlin-Heidelberg-New York,1981.

[7] G.W.Gibbons and S.W.Hawking, Phys.Rev.D15,2752-2756 (1977).

[8] S.W.Hawking and W.Israel, eds.,General Relativity, An Einstein Centenary Survey, Cambridge University Press, Cambridge, 1979.

[9] S.S.Chern, S.Smale eds.,Global analysis, Proceedings of Symposia in Pure Mathematics Vol.XV, American Mathematical Society, Providence, Rhode Island, 1970.

[10] J.A.Wheeler in Relativity, Groups and Topology, eds, C.De Witt and B.De Witt, Gordon and Breach, New York, 1964.

[11] N.H.Kuiper, Jber. Deutsch.Math. Verein 69,77-88 (1967).

[12] T.Banchoff, J.Diff.Geom.1,245-256 (1967).

[13] R.Conelly, in Proceedings of the International Congress of Mathematics, Helsinki,1978.

[14] N.H.Kuiper, in Chern Symposium 1979, Springer Verlag, New York-Heidelberg-Berlin, 1980.

[15] J.Cheeger, Proceedings of the National Academy of Sciences,USA,76,No.5,2103-06(1979).

[16] W.Müller, Adv.in Math. 28,233-305 (1978).

[17] D.Stone, Indiana Univ.Math.J.28,1-21(1979).

[18] J.Cheeger, in Proc.Symp.Pure Math.,Vol36, R.Osserman, A.Weinstein, eds., Providence, Rhode Island, 1980.

[19] J.Cheeger, Spectral geometry of singular riemannian spaces, to appear.

[20] I.M.Singer and J.A.Thorpe, Lecture Notes in Elementary Topology and Geometry, Scott, Foresman and Co, Glenview, Ill. 1967.

[21] J.A.Wheeler, private communication.

[22] J.R.Munkres, Elementary Differential Topology, Princeton University Press, Princeton, N.J.1966.

[23] S.S.Chern, J.Math.and Mech. 16, 101-118 (1966).

[24] L.A.Santaló, Integral Geometry and Geometric Probability, Addison Wesley, London 1976.

[25] H.Weyl, Amer.J.Math. 61,461-472 (1939).

[26] P.Wintgen, Normal cycle and Integral Curvature for Polyhedra in Riemannian manifolds, Colloquia Mathematica Societatis János Bolyai, Budapest, 1978.

[27] J.Cheeger, unpublished notes (1977).

[28] K.G.Wilson, Phys.Rev.D10,2445-2459 (1974).

[29] F.J.Wegner, J.Math.Phys.12,22-59 (1971).

[30] J. Cheeger, A. Karpf, W. Müller, R. Schrader, E. Tränkle, unpublished.

[31] D.Ruelle, Statistical Mechanics, Benjamin, New York, 1969.

[32] K.Osterwalder and R.Schrader, Comm.Math.Phys.31,83-112 (1973),42,281-305 (1975).

[33] A.D. Alexandrov, Die innere Geometrie der konvexen Flächen, Akademie Verlag, Berlin (1955).

[34] A.W. Pogorelov, Regularity of a convex surface with given Gaussian curvature, Mat. Sb. 31 (73) 88-103 (1952).

[35] A.W. Pogorelov, Extrinsic geometry of convex surfaces, Transl. Math. Monographs (1973).

PARTICLES AND GEOMETRY

John Archibald Wheeler
Center for Theoretical Physics
Department of Physics
University of Texas
Austin, Texas 78712 U.S.A.

Heisenberg's Search for Wholeness in Physics

One who was privileged at the age of twenty-three to walk and talk with Heisenberg about nuclear physics in Copenhagen, one who was a fellow lecturer with him at Ann Arbor in 1939, one who had warm and interesting discussions with him on general relativity in his office in München in 1966, one who saw him and his wife in the United States in 1973 giving their all to the representing of German culture in the great tradition once again in the wider world thinks of him as present with us in spirit today.

Concentrate as Heisenberg did throughout his life first on one great issue then another and then another, he had also a special gift for seeing physics in its wholeness. At that moment when he and Dirac were pushing forward electron-positron pair theory to the most distant frontiers yet attained, Heisenberg had the judgment to realize that the electron theory was not the final answer. No one saw earlier than he that nuclear matter had to be built of neutrons and protons. Subsequently, when he was pushing forward the theory of nuclear matter, he further recognized the role that non-nucleonic entities must play in the binding of nucleons. Still later, at a time when baryons and mesons seemed the center of nuclear physics, Heisenberg had the perspective to concern himself with a wider view of the family of particles. Subsequently, when particle physics in this wider sense was in its first bloom, he could discourse knowingly and enthusiastically on the prospects opened up by Einstein's vision of gravitation, not as a "foreign" field transmitted through space, but as a manifestation of space curvature itself.

Today we see the reuniting of those two ostensibly different ways of exploring physics represented by the two words "particles" and "geometry". What we call "supersymmetry" plays a leading part in drawing particles and geometry into this larger unity. No one would be happier than Heisenberg at this development. But again today, as so many times in the past, would he not direct our attention to some still further decisive consideration needed for wholeness?

The Mystery of the Quantum

Particles and geometry plus something, yes; but particles and geometry plus *what*? Plus, surely, some consideration of the most mysterious element in all of physics, the quantum. No topic gripped Heisenberg like the quantum from his youth to his last days. None looks more like being the central overarching principle of twentieth century physics. It is vehicle for all we do. Yet we are as lost as ever in fathoming its deep and secret underpinnings. The bicycle too is a useful vehicle. However, the minute we ask what holds it up, we fall over. Do we fear a similar fate as we "ride" the quantum? Is it not rather the mystery of the quantum that deters us from asking the origin of the quantum?

It has taken all the encouragement and judgment we could muster over the years to analyze particles and fields of force and even the dynamics of space geometry itself. These subjects in the long ago belonged to philosophy. Little by little physics has seen that philosophy is too important to be abandoned to the philosophers. But the central core of philosophy, the mystery of existence itself, has until now lain "off limits". Is it possible that the nature of existence can only be understood when we understand the origin of the quantum? Is it possible that we can understand the origin of the quantum only when we understand the nature of existence?

To run away from these questions can hardly help in answering them. On the contrary, where better than at this meeting can we start to ask these questions? Where more inspiringly than here does one see the gathering unity of physics? Where better than in this München conference can one take seriously Heisenberg's counsel to look from the part to the whole--von der Teil zu dem Ganze (Heisenberg, 1969)?

Einstein's Vision of Geometry

We attribute to Einstein the achievement of making space geometry a part of physics, coordinate with the electromagnetic field of Faraday and Maxwell and with the fields receiving attention here in binding mass to mass. Einstein in turn attributed the inspiration for this development to three sources: Riemann, Mach and 1908.

Riemann taught that geometry is part of physics. As Einstein (1934) put it, "...space was still, for them [physicists], a rigid homogeneous something, susceptible of no change or conditions. Only the genius of Riemann, solitary and uncomprehended, had already won its way by the middle of the last century to a new conception of space, in which space was deprived of its rigidity, and in which its power to take part in physical events was recognized as possible."

Mach taught that *mass*--and he was speaking of the masses of the distant stars--"there" determines, directly or indirectly, the conditions of motion "here". Einstein spoke of his own geometric theory of gravity as a working out of Mach's program (Einstein, 1913; Harneck, 1979; Isenberg and Wheeler, 1979), in the sense that space curvature "there" conditions space geometry "here".

These were two of the intellectual antecedents of the geometric theory of gravitation. Ludwig Boltzmann was familiar with both and tried to bring them together (original lecture in the mid-1890's, Boltzmann, 1904) but felt he could not: "...the separations of the fixed stars perhaps can only be represented in a non-Euclidean space of extremely small curvature, which would indeed also make a connection with the law of inertia to this extent, that then a freely moving object would have to return in eons to its original position in case the curvature is positive."

Boltzmann failed to go further. Einstein succeeded. He linked Riemann's "geometry is physics" and Mach's "inertia here is conditioned by mass there" by his discovery of the equivalence principle in 1908

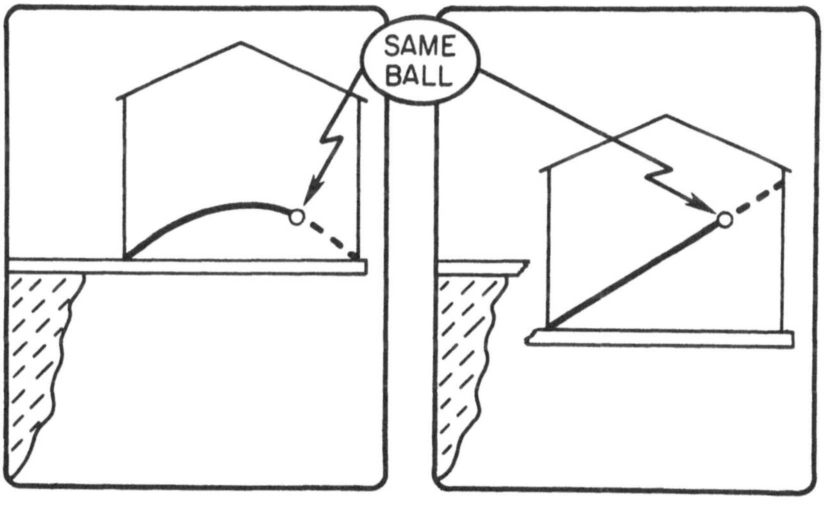

Figure 1. Gravity, evident in the everyday frame of reference at the left, disappears in the free float frame of reference at the right. The ball is thrown from the same place in the same direction with the same velocity in both. The difference in appearance of the track in the two cases arises, not from the ball, but from the frame of reference from which it is regarded. The "fault" of the "force on the ball" is not the Earth but the room.

(as described by Einstein, 1949). In brief, Einstein's first step in explaining gravity was "to take away gravity". He taught us to describe motion in a free float frame of reference (Figure 1). He emphasized that physics only then lends itself to simple description when it deals with *local* quantities. Gravitation, apparently anulled, came back into evidence (Figure 2) in the relative acceleration of nearby test masses. This feature of nearby worldlines Einstein identified with space curvature. With this concept in hand Einstein was well on the way to his 1915 and still standard gravitation theory or "geometrodynamics". Today we put it in these words, "Space tells mass how to move. Mass tells space how to curve."

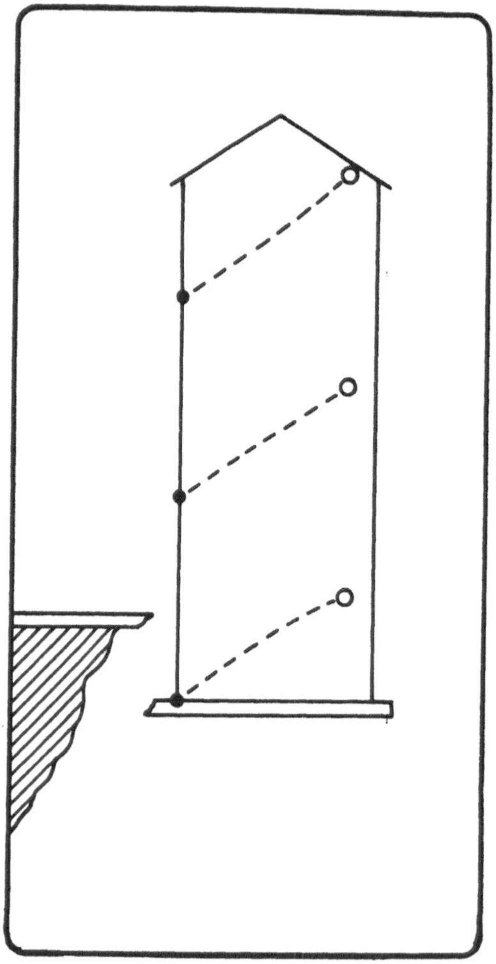

Figure 2. Relative acceleration of nearby text particles as the local measure of gravitation with which Einstein replaced Newton's action at a distance.

Illusion and Reality

Einstein's theory went farther than previous physics in picking our sensations to pieces. "Sound" we had learned is really pressure waves; "light", really electromagnetic waves. But the arc of the ball after it is thrown from the hand (Figure 1), is not that at least real? Not even that. Nowhere more dramatically than in gravitation physics do we see the age-old destiny of physics. Physics is a magic window. It shows us the illusion that lies behind reality--and the reality that lies behind illusion. In our own day in chromodynamics we see the interplay of illusion and reality, of theoretical substructure and observational finding in a more sophisticated form than ever before.

The Implications of Heisenberg's Principle of Indeterminism for Field Dynamics

It took long to recognize that geometrodynamics, like electrodynamics, admits and demands the "great division", the division between the dynamic law itself and the initial conditions. Today, thanks to the work of many distinguished investigators, we know how to express the dynamics of geometry in Hamiltonian form, and how to state the initial conditions in terms of "field coordinates" and "field momenta".

Both field coordinate and field momentum are needed to predict the future. However, exactly that demand Heisenberg's principle of indeterminism rules out as impossible. We were prepared for this lesson and its consequences by the wonderful development of quantum electrodynamics; and even before that by the quantum mechanics of the elementary harmonic oscillator. There too we could not know both the initial momentum and the initial coordinate; there too we could not predict the future in all detail. In that simpler context, however, we learned that every physically sensible question permits a sensible answer.

In no way do the simplicities of the quantum mechanical description show up more clearly than in the familiar expression for the wave function for the ground state of the harmonic oscillator,

$$\psi(x) = N \exp(-m\omega x^2/2\hbar), \tag{1}$$

in the coordinate representation; or in the momentum representation,

$$\phi(p) = N' \exp(-p^2/2m\omega\hbar). \tag{2}$$

It is a small step, we know, from the ground state of one oscillator to the ground state of many oscillators, and from there to the ground state of the electromagnetic field in the *spacelike representation*,

$$\psi(\mathbf{B}(x,y,z)) = N \exp\left\{-\iint \frac{\mathbf{B}(x_1)\cdot\mathbf{B}(x_2)}{16\pi^3\hbar c r_{12}^2}\, d^3x_1 d^3x_2\right\}. \qquad (3)$$

Ours is the choice whether this representation shall be stated in terms of the field coordinate--here the magnetic field--or the field momentum, the electric field, likewise expressed in its dependence on position in space: $\mathbf{E}(x,y,z)$ (expression for ψ formally identical with (3) except for the replacement of $\mathbf{B}$ by $\mathbf{E}$).

Long ago S. Tomonaga (1946) taught us his "bubble differentiation" and his wave equation for the alteration of the quantum mechanical probability amplitude, ψ, in its spacelike representation with the advance of "many-fingered time". The general character of Tomonaga's equation is not greatly altered, we know, if one turns attention from the probability amplitude itself to the classical Hamilton-Jacobi function, actually a functional, $S(\mathbf{B}(x,y,z))$, that gives the phase of this wave function,

$$\psi = (\text{slowly varying factor}) \exp(iS/\hbar). \qquad (4)$$

More recently Asher Peres (1962) gave us the corresponding Hamilton-Jacobi equation for Einstein's standard geometrodynamics,

$$g^{-1/2}\left(\frac{1}{2}\, g_{pq}g_{rs} - g_{pr}g_{qs}\right)\frac{\delta S}{\delta g_{pq}}\,\frac{\delta S}{\delta g_{rs}} + g^{1/2}R = 0. \qquad (5)$$

This equation summarizes more succinctly than any other description we possess the entire story of the Hamiltonian dynamics of empty space. It is the analog of the Hamilton-Jacobi equation for a free particle in relativistic mechanics,

$$(\partial S/\partial t)^2 - (\nabla S)^2 - m^2 = 0. \qquad (6)$$

The "Square Root of Gravity"

Dirac (1928a, 1928b) taught us how to take the square root of this equation or the corresponding Klein-Gordon equation. Claudio Teitelboim (1977) and R. Tabensky and Teitelboim (1977) showed us how to "take the square root of general relativity"--in the sense of equation (5)--and how beautiful and natural the mathematics is and how faithfully one reproduces in this quite independent way the theory of supergravity that had been given us the year before by Freedman, van Nieuwenhuizen and Ferrara (1976) on the one hand and by Deser and Zumino (1976) on the other.

Do we worry that supergravity with its spinor character will assign to a completely collapsed object, a black hole, some "hair", some attribute other than those three properties, mass, charge and angular momentum, which alone, we have been taught, remain after all the original details have fallen away? We might as well worry, Cordero and Teitelboim (1978) and Cordero (post 1978) tell us, about thinking of the momentum, p, of a black hole in the Christodoulou formula (Christodoulou, 1970),

$$E^2 = M_{ir}^2 + \frac{J^2}{4M_{ir}^2} + p^2, \tag{7}$$

being some new attribute of the object. In actuality it is a feature, we know, not of the object, but of the frame of reference from which we regard it. Likewise, the most general black hole with spin 3/2 supercharge is equivalent, under supersymmetry transformation, to a black hole without supercharge.

It never ceases to amaze that the supersymmetry forced on us by the rich data and profound investigations of elementary particle physics is identical in so many respects with the supersymmetry that we get directly out of general relativity plus quantum mechanics. In this wonderful new connection between two great domains of investigation we surely have something deep and precious.

Quantum Fluctuations--Indication That Time Cannot Be A Primordial Category In The Description of Nature

Another connection between particles and geometry we see in fluctuations: fluctuations in the chromodynamic field at small distances as described by Richard Feynman (1981) and Kenneth Johnson (1981); and fluctuations in geometry at small distances. Already from electrodynamics we knew, to use the word of John Klauder, that "unruly" configurations predominate (Bohr and Rosenfeld, 1933). This feature common to all fields, shows up particularly clearly in expression (3) for the ground state functional of the electromagnetic field in the spacelike representation. It has no sense to speak of the probability for the magnetic field, $\underset{\sim}{B}$, to assume this, that or the other value at a point. Even when we ask at the hands of (3) for the probability of a magnetic field of value $\underset{\sim}{B}$ over a spacelike region of extension L, we learn that all values of B have comparable probability up to a value of the order

$$\Delta B \sim (\hbar c)^{1/2}/L^2, \tag{8}$$

which is the larger, the smaller the region.

In the case of geometry similar considerations tell us that the inescapable quantum fluctuations in space curvature, R, in a region of extension L are of the order of magnitude

$$\Delta R \sim L^*/L^3, \tag{9}$$

where $L^* = (\hbar G/c^3) = 1.6 \times 10^{-33}$ is the Planck length (Wheeler, 1957). In other words, space is to be compared with the ocean which looks smooth to the air traveller 10 km above it, shows waves when one descends to 100 meters above it and is seen to be covered with foam when one is in a life raft right on the surface. Hermann Weyl, discussing multiply-connected geometry in 1927 (Weyl, 1949), remarked, "...a more detailed scrutiny of a surface might disclose that, what we had considered an elementary piece, in reality has tiny handles attached to it which change the connectivity character of the piece, and that a microscope of ever greater magnification would reveal ever new topological complications of this type, *ad infinitum*." Today it is not geometric fancy but quantum compulsion that leads us to think of space geometry as multiply connected in the small and everywhere endowed with a foamlike structure (Wheeler, 1957; Wheeler, 1968, figure 5; Gibbons and Hawking, 1977; Perry, Hawking and Gibbons, 1978).

There was a time when to take seriously all-pervasive quantum fluctuations even in the electromagnetic field took uncommon imagination. Yet faith in the reality of those fluctuations did as much as anything to open the door to the Lamb shift, to renormalization and to the unprecedented development of quantum electrodynamics following World War II. Quantum fluctuations in the geometry and connectivity of space at small distances surely have deep lessons of their own to teach.

Superspace as Way to See the Consequences of Fluctuations in Geometry

Among alternative ways to look at the predicted fluctuations in geometry and their consequences for the concept of "time" one of the most instructive is provided by the superspace picture of the dynamics of geometry (Wheeler, 1968) (Figure 3). In that picture one sees impressively the great difference in principle between spacetime--the classical history of space geometry evolving deterministically with many-fingered time--and the quantum description. There are several alternative and equivalent formalisms for describing the quantum

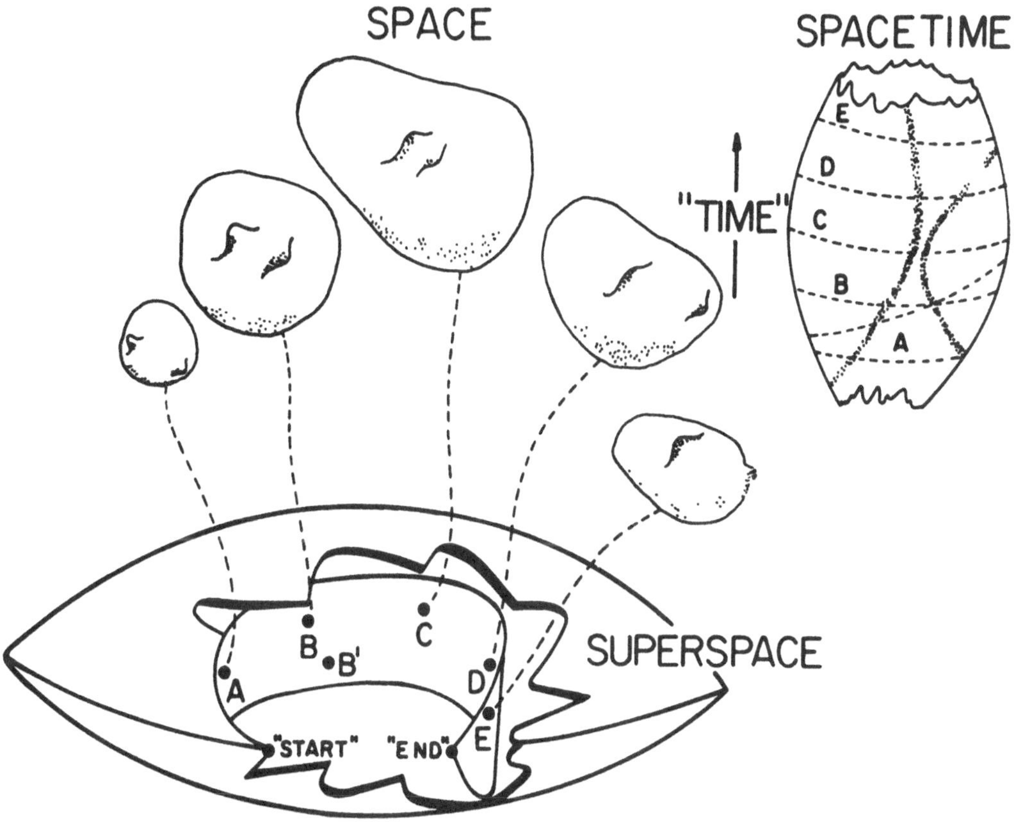

Figure 3. Space (five configurations in the upper left), spacetime (upper right) and superspace (below). Each point in this infinite-dimensional superspace of geometrodynamics represents a complete 3-dimensional geometry in all its detail. The 3-dimensional geometries obtained by making this, that and the other spacelike slice through a given classical spacetime do not fill out all of superspace. They fill out a maniforld in superspace which, although infinite-dimensional, has only 1/3 of the dimensionality of superspace itself: not a "worldline" but a *"leaf of history"*, cutting through superspace as indicated in the diagram. Thus there is a sharp distinction between the "yes" 3-geometries that occur in a given history of geometry evolving deterministically with many-fingered time and the "no" 3-geometries that don't. In quantum theory, by contrast, there is no such sharp "yes, no" distinction. The 3-geometries that occur with appreciable probability amplitude spread out beyond the infinitesimally thin classical leaf of history. They are far too numerous to be accomodated in any 4-dimensional spacetime. Legalistically speaking there is no such thing as spacetime.

dynamics of geometry. In a superspace description one speaks of the probability amplitude for this, that or the other 3-geometry, as in equation (3) one speaks of the probability amplitude for this, that or the other magnetic field configuration, or in (1) for this, that or the other oscillator coordinate. It is wrong to ascribe a worldline to the harmonic oscillator. It is equally wrong to ascribe a deterministic history to the electromagnetic field or to geometry. Spacetime is not a legitimate concept, according to quantum theory. It collides head-on with the battle-tested demands of Heisenberg's principle of indeterminism.

Under everyday circumstances and in atomic physics and even at the distances of $\sim 10^{-16}$cm probed in high energy experiments one is so far away from the Planck scale of distances that the concept of spacetime preserves its validity, we are prepared to believe, to a fantastic level of precision. At still smaller distances do all concepts of space and time fail for some as yet completely unappreciated reason? We have as yet no way to know. However, quantum theory assures us that the concept of spacetime certainly does fail when we get down to 10^{-33}cm. Then even the very ideas of "before" and "after" lose all significance.

One does not have to devise some unimaginable machine to encounter the very smallest of distances. Nature itself in the shape of big bang and gravitational collapse makes conditions more extreme by far than man can hope to achieve. Nowhere more than at these two limits of time can one believe that the terms "before" and "after" lose all relevance and application.

The Austerity Hypothesis

Einstein's field equation as applied to classical cosmology never gave the slightest warrant for a "cycle after cycle" universe. Still less does quantum theory provide any reason whatsoever to think of a "before" before the big bang or an "after" after collapse. On the contrary, we are invited today more strongly than ever to believe that the laws of physics could not have stood chiseled upon a tablet of granite from everlasting to everlasting (Figure 4). Down at the bottom of things there could not have been any prefabricated gears and pinions, any trained assembly crew, to put it all together. No way out is evident except the *austerity hypothesis*: that the laws of physics came into being out of nothing, and this by a process as unplanned and higgledy-piggledy as genetic mutation or the second law of thermodynamics.

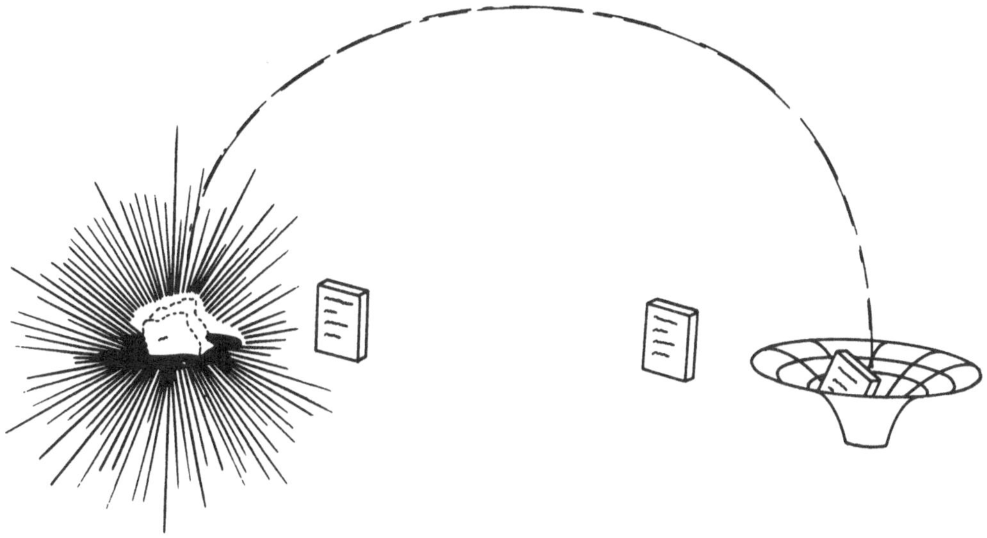

Figure 4. The laws of physics could not have endured from "everlasting to everlasting". They are symbolized here as coming into being in the Big Bang and fading out of existence in gravitational collapse.

The universe, according to the austerity hypothesis, has to construct part without part, fields without fields, spacetime without spacetime and law without law.

To uncover the great regularities of physics demanded decades of labor by hundreds of gifted investigators and thousands of experiments. What could be more disillusioning--and at first sight more unwelcome--than to think of all that beauty founded upon pure chaos? More like being thrown into cold water? But in the end what could be more invigorating, more inspiring, more challenging, more productive of guidance than the working hypothesis that everything has to come into being out of nothing?

Elementary Quantum Phenomenon as Elementary Building Unit

No other building unit offers itself except one that is itself neither a particle nor a field nor a spacetime, but one that requires all three for its action; one that has no weight, exerts no force, has no locale, but lies at the center of the central theme of modern physics: the *elementary quantum phenomenon* (Figure 5).

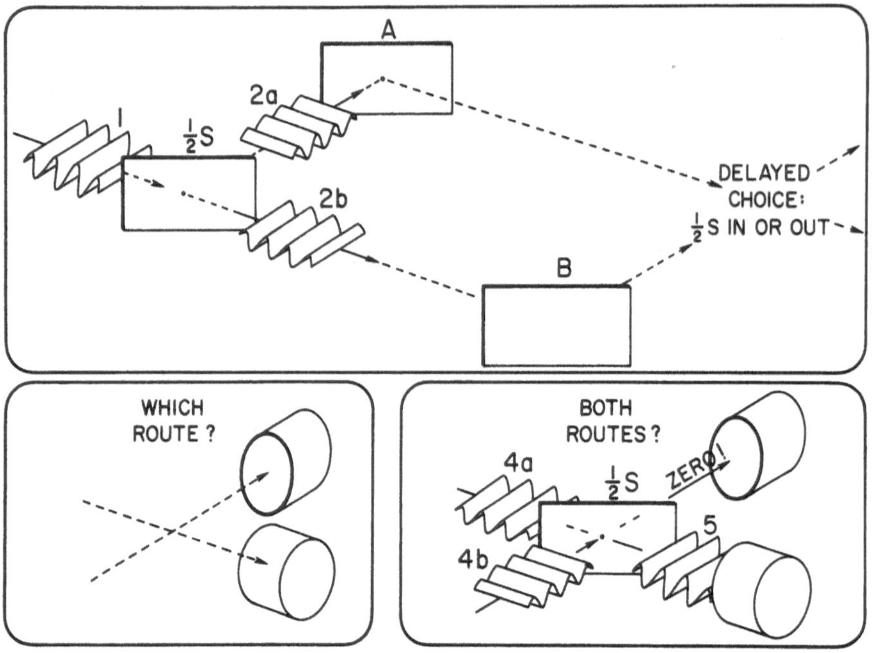

Figure 5. *Elementary quantum phenomenon* illustrated in the context of the beam splitter (above) as used to perform a delayed-choice experiment (below). An electromagnetic wave comes in at I and encounters the half-silvered mirror marked "½S which splits it into two beams, 2a and 2b, of equal intensity which are reflected by mirrors A and B to a crossing point at the right. Counters (lower left) located past the point of crossing tell by which route an arriving photon has come. In the alternative arrangement at the lower right, a half-silvered mirror is inserted at the point of crossing. On one side it brings beams 4a and 4b into destructive interference, so that the counter located on that side never registers anything. On the other side the beams are brought into constructive interference to reconstitute a beam, 5, of the original strength, 1. Every photon that enters at 1 is registered in that second counter in the idealized case of perfect mirrors and 100% photodetector efficiency. It might seem that in the one arrangement (lower left) one finds out by *which* route the photon came.

(Figure 5, continued) It might seem that in the other arrangement (lower right) one has evidence that the arriving photon came by both routes. In the new "delayed-choice" version of the experiment [Early version of such a concept put forward by Weizsäcker (1931) at age 18, inspired, he tells us, by Heisenberg, four years before the Einstein-Podolsky-Rosen experiment; exchange of letters between Weizsäcker and Jammer relating these considerations to the Einstein-Podolsky-Rosen experiment quoted in Jammer (1974); terminology "delayed-choice" and delayed choice experiments generally in Wheeler (1978).] one decides whether to put in the half-silvered mirror or take it out at the very last minute. Thus it might seem that one decides whether the photon "shall have come by one route, or by both routes "after it has *already done* its travel". However it is the great lesson of the Bohr-Einstein dialog that one has no right even to speak of the "route" of the photon in all its long travel [here a meter, but billions of lightyears for the lensed quasar 0957 + 561A, B (Wheeler, 1981)] from the point of emission to the point of reception. No elementary quantum phenomenon is a phenomenon until the phenomenon is recorded (registered or "observed" or "brought to a close by an irreversible act of amplification").

No other method of construction suggests itself except one that is (1) reflective and (2) statistical. Particles, fields and spacetime furnish the stage setting for individual quantum phenomena. But do quantum phenomena in turn, by the billion, build the evidence for, and in that sense constitute, all that is? We would seem to be compelled to ask this question if we take seriously the austerity hypothesis.

The questions we raise here are far beyond today's power to answer. One who troubles about them, however, would be wanting in responsibility if he did not draw attention to three remarkable features of physics that have the flavor of austerity. One is the amazing ease of deriving the Einstein dynamics of geometry and likewise the modern dynamics of the color field, out of the fantastically simple requirement that that dynamics should be embeddable in a spacetime manifold (Hojman, Kukar and Teitelboim, 1973, 1976; Teitelboim, 1973a,b, 1976; Kuchar, 1973, 1974; briefly summarized and discussed in Wheeler, 1980).

Two further indications suggest almost everything comes from almost nothing. A guiding principle of algebraic geometry, "the boundary of a boundary is zero" occupies a central position in electromagnetism, gravitation and chromodynamics. That consideration lets left, the quantity that appears is not the full Riemann curvature tensor but the Einstein curvature tensor of which typical components are

$$\begin{aligned} G^0{}_0 &= -(R^{12}{}_{12} + R^{23}{}_{23} + R^{31}{}_{31}), \\ G^1{}_1 &= -(R^{02}{}_{02} + R^{03}{}_{03} + R^{23}{}_{23}), \\ G^0{}_1 &= R^{02}{}_{12} + R^{03}{}_{13}, \\ G^1{}_2 &= R^{10}{}_{20} + R^{13}{}_{23}. \end{aligned} \qquad (15)$$

What is the meaning of these strange equations?

Source and Field? Which the Slave, Which the Master?

In gravitation as well as in electromagnetism we used to look at the source as primary and the field as secondary. The source (electric charge or mass energy) "knew" that it wanted to be conserved and the field ran along behind as its slave, obedient to its wish. Today we regard the field as primary and the source as secondary. Without the field to govern it, the source would not know what to do. It would not even exist. When two gigantic spaceships smash into each other, much is destroyed. One quantity, we know, is conserved, the energy momentum 4-vector. What master is so powerful that it can hold those two mighty spaceships in straightline motion before they hit and see to the conservation law in the crash itself? Space! Space grips them both. Space right where they are enforces the conservation of momentum and energy. (Figure 7.)

We used to think of the universe, figuratively speaking, as made out of gears and pinions and put together with the sophistication of an elaborate watch. Today what we know of the big bang and gravitational collapse encourages us to think just the opposite. We see no other way for the laws of physics to come into being except higgledy-piggledy chance.

The "austerity hypothesis" suggests to us that almost everything comes into being out of almost nothing. Of "almost nothing" principles it is difficult to think of one more natural than the central principle of algebraic topology: the boundary of a boundary is zero, $\partial\partial = 0$. (Figure 8)

The principle $\partial\partial = 0$ lies at the very heart of electrodynamics, geometrodynamics and chromodynamics. In each field theory it occurs twice, once at the 1-2-3-dimensional level, and again at the 2-3-4-dimensional level.

itself be spelled out most simply at the level of a classical field theory but is equally germane in each case to the corresponding quantum field theory. The other consideration has to do exclusively with the quantum. Wooters (1980, 1981) formulates a striking connection between quantum mechanics and statistics: The distinguishability between two states, defined in the purely statistical sense of distinguishability, is measured by angle in Hilbert space. Both "the boundary of a boundary" and "distinguishability" are sufficiently interesting in themselves, and also sufficiently illustrative of what one means by "austerity" as to deserve a little spelling out.

The Boundary of a Boundary is Zero

What does it mean to say that space tells mass how to move and mass tells space how to curve? Energy can be considered to be the decisive quantity in governing the action of mass on geometry. But the energy of a system of particles and fields is only one component of the 4-vector, p, of energy and momentum. The amount of energy depends both on the observer and on the region observed:

$$E = u^{\alpha} T_{\alpha}{}^{\beta} d\Sigma_{\beta} \ . \tag{10}$$

This equation defines the stress energy tensor, $T_{\alpha}{}^{\beta}$, the source term in Einstein's field equation. There is a great difference between electromagnetism and gravity. In the case of electromagnetism we speak about the force on one particle,

$$D^2 x^{\alpha}/D\tau^2 = (e/m) F^{\alpha}{}_{\beta} (dx^{\beta}/d\tau). \tag{11}$$

In the case of gravitation there is no force on one particle:

$$D^2 x^{\alpha}/D\tau^2 = 0. \tag{12}$$

The physics name for Einstein's local measure of gravitation, we know, is "tide-producing acceleration"; and the mathematics name is "geodesic deviation" or "spacetime curvature".

The separation, η^{α}, between two nearby test masses is given in terms of the curvature of spacetime, $R^{\alpha}{}_{\beta\mu\nu}$, by the "equation of geodesic deviation",

$$D^2 \eta^{\alpha}/D\tau^2 = -R^{\alpha}{}_{\beta\mu\nu} (dx^{\beta}/d\tau) \eta^{\mu} (dx^{\nu}/d\tau). \tag{13}$$

This equation is as important for gravitation as is the Lorentz equation for electromagnetism. It opens the door to experimental methods for measuring spacetime curvature.

It takes considerable thinking to see how to get each component of the curvature individually by measurements of the relative separations of nearby test particles. However, in the end, the language of physics is equivalent to the language of geometry. In geometric language, the curvature is measured by the rotation of a vector transported parallel to itself around a closed curve (Figure 6).

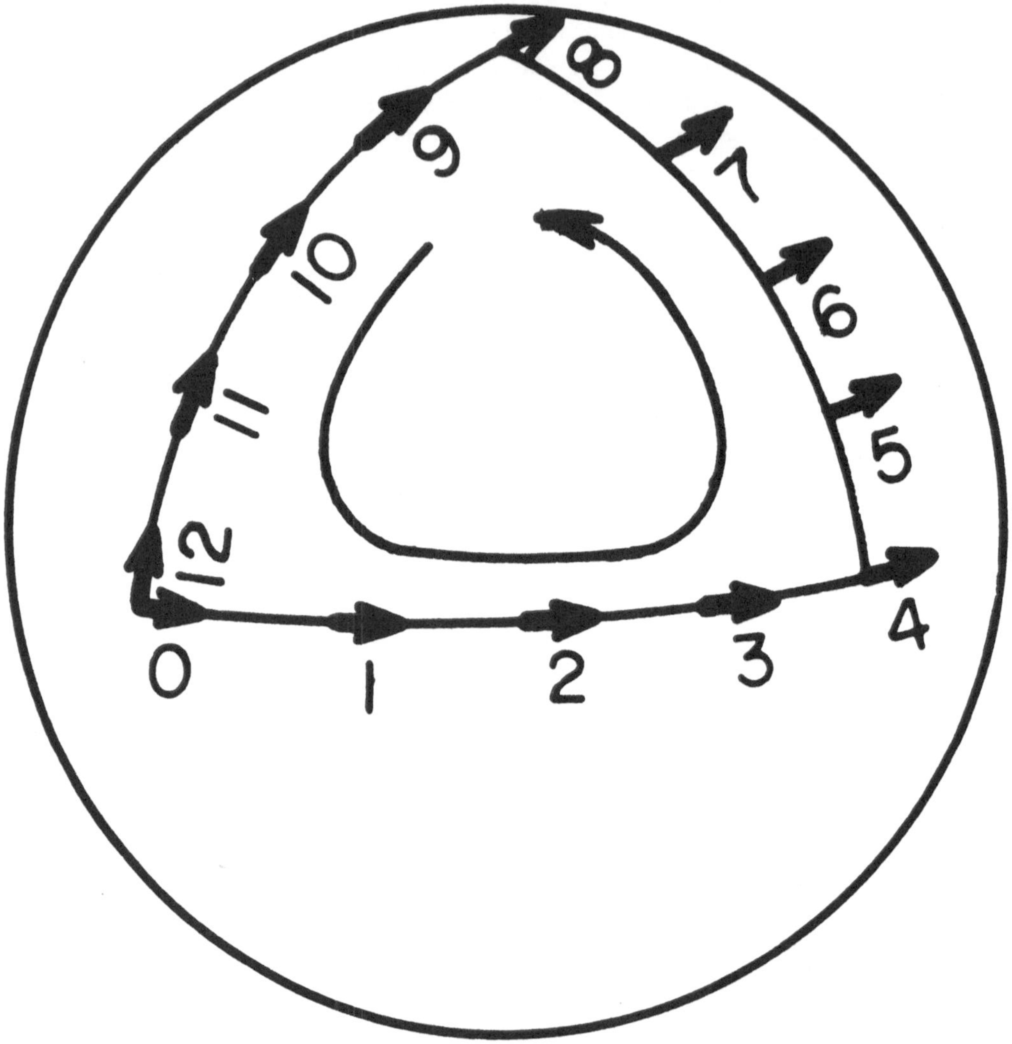

Figure 6. The vector transported parallel to itself around the indicated closed circuit undergoes a rotation of 90°. The curvature is (rotation of vector)/(area circumnavigated) = $(\pi/2/(4\pi a^2/8) = 1/a^2$, where a is the rediuo of the sphere.

Einstein's field equation reads

$$G_{\alpha}{}^{\beta} = 8\pi T_{\alpha}{}^{\beta}. \tag{14}$$

Here the quantity on the right is the stress energy tensor. On the

Figure 7. Space grips mass, keeps it going straight when free and by its power enforces conservation of energy and momentum in a smash. The coupling of mass and geometry, far from being the weakest force in nature, is the strongest.

The 1-dimensional boundary of one face of the cube is the collection of four directed segments or edges. The 2-dimensional boundary is the collection of all six faces. When we add up the four edges for each of the six faces, or twenty-four edges altogether, we get zero because the edges cancel in pairs. Thus the 1-dimensional boundary of the 2-dimensional boundary of the 3-diminsional cube is zero: $\partial\partial = 0$.

The idea $\partial\partial = 0$ applies to curvature. We subject a vector in imagination to parallel transport around one face of a little cube in spacetime. It undergoes a rotation. However, the sum of the rotations associated with all six faces is zero. "The rotation diagram closes."

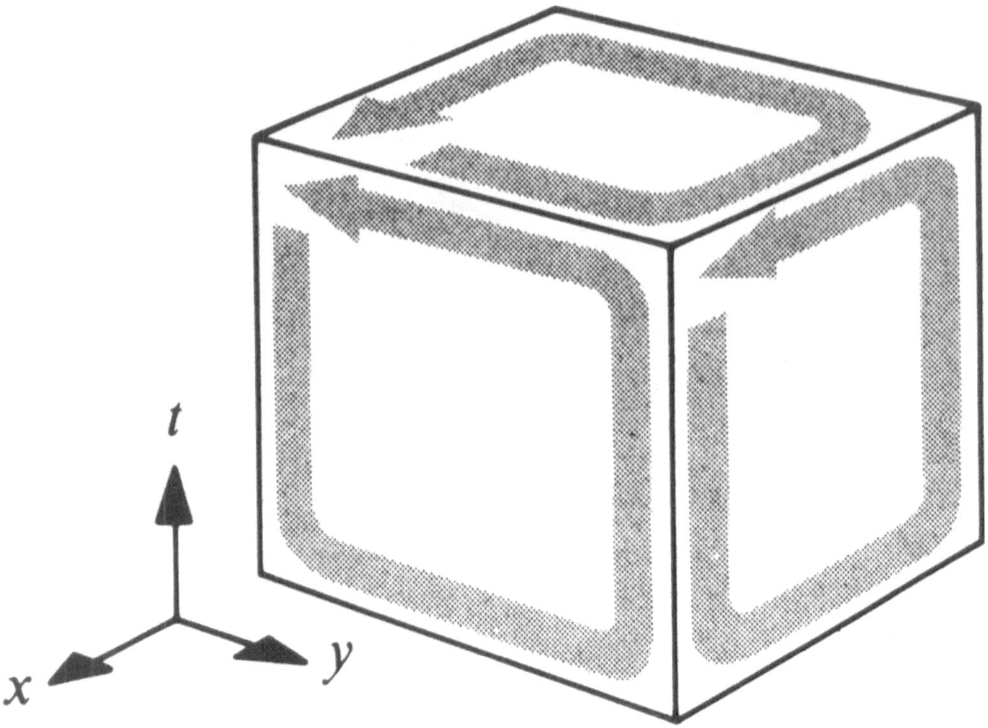

Figure 8. The 1-dimensional boundary of the 2-dimensional boundary of the 3-dimensional cube is zero.

The alteration in the vector V^{α} in parallel transportation around the front face of the cube is

$$\delta V^{\alpha} = -R^{\alpha\beta}{}_{31}(x,y+\tfrac{1}{2}\Delta y,z)V_{\beta}\Delta x^{3}\Delta x^{1}. \qquad (16)$$

The combination of the effects on V^{α} produced by going around the front and the back face is

$$-(\partial R^{\alpha\beta}{}_{31}/\partial x^{2})V_{\beta}\Delta x^{1}\Delta x^{2}\Delta x^{3}. \qquad (17)$$

The vanishing of the totalized rotation for all six faces gives the so-called full Bianchi identity,

$$(\partial R^{\alpha\beta}{}_{23}/\partial x^{1}) + (\partial R^{\alpha\beta}{}_{31}/\partial x^{2}) + (\partial R^{\alpha\beta}{}_{12}/\partial x^{3}) = 0. \qquad (18)$$

How are we to "wire up" stress energy, our source, to spacetime geometry, our field, so as to guarantee conservation of the source, and do this automatically, without any "gears and pinions and watch-works"? By applying the principle $\partial\partial = 0$ at the 2-3-4-dimensional level.

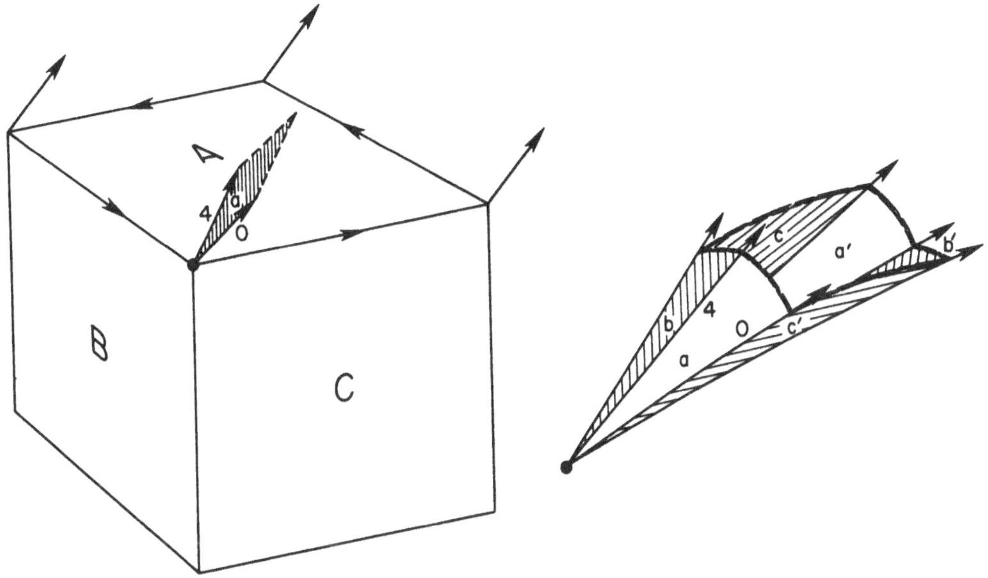

Figure 9. Rotations and the 1-boundary of the 2-boundary of a 3-cube. Left: The rotation a experienced by a sample vector on being transported parallel to itself around the circuit 0,1,2,3,4. This rotation measures one component of the spacetime curvature or gravitational tide-producing acceleration. Right: The rotations associated with all six faces together add up to zero; the diagram closes. It closes because each edge of the cube is traversed twice, and in opposite directions in the circumnavigation of the two abutting faces of the cube: $\partial\partial = 0$.

"No Creation of Source"

We consider a region of spacetime, a 4-cube $\Delta x\Delta y\Delta z\Delta t$. We demand that there shall be no creation of source in the given 3-dimensional region in the given time. That means that the amount of energy-momentum in the 3-cube $\Delta x\Delta y\Delta z$ at the time $t + \frac{1}{2}\Delta t$ is to be equal to the amount in the same region at the time $t - \frac{1}{2}\Delta t$ with corrections for what flows across the six surfaces ($\Delta x\Delta y\Delta t$ at $z + \frac{1}{2}\Delta z$ and five others) in the intervening time and with no other corrections. (Figure 10) We want zero creation of source in the 4-cube $\Delta x\Delta y\Delta z\Delta t$. That means a zero total for the energy-momentum in the eight 3-cubes that bound that 4-cube. If this result is to come about automatically, via $\partial\partial = 0$, then the energy momentum in any given 3-cube must in turn be the sum of contributions from each of the six 2-faces of that cube.

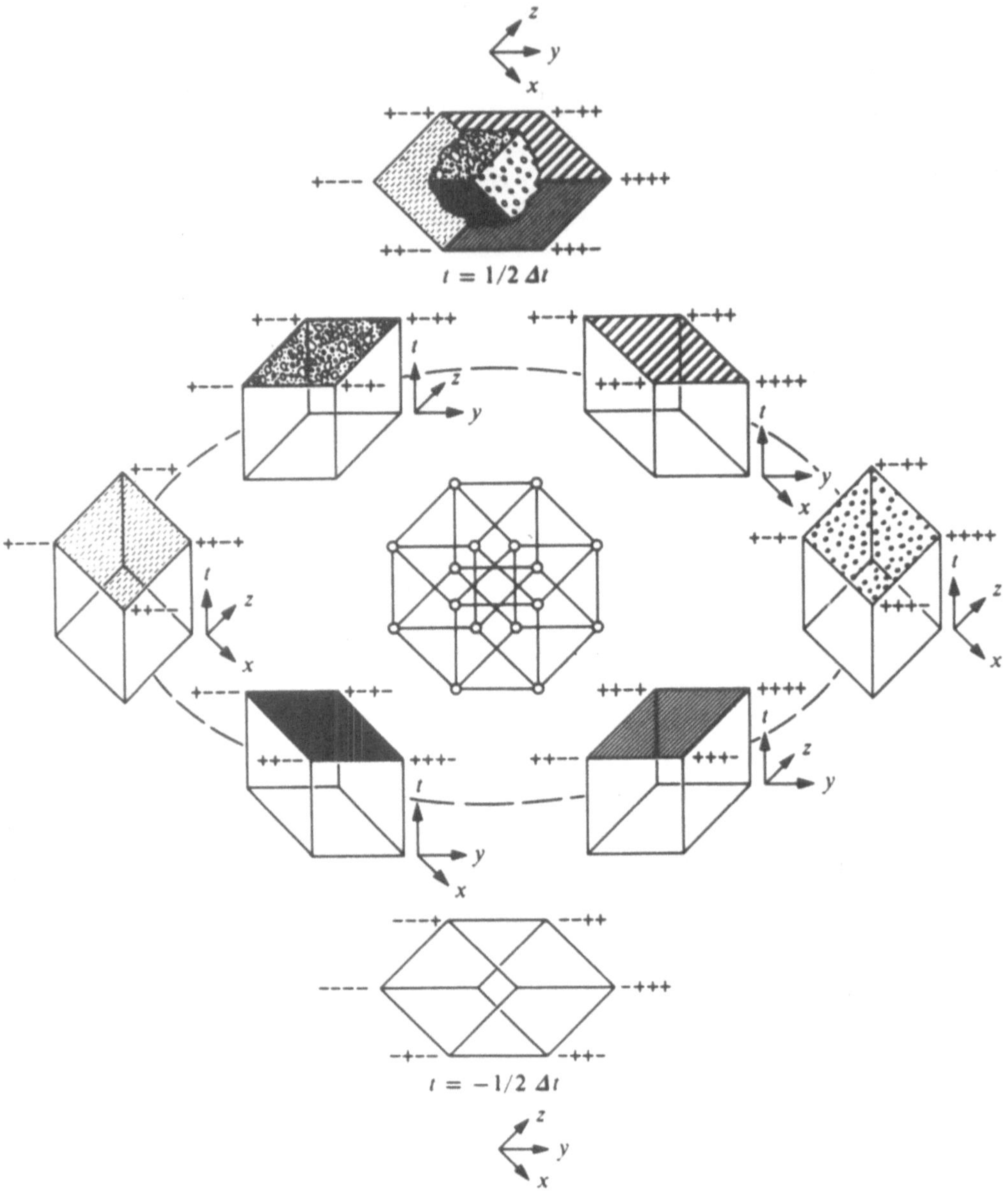

Figure 10. The 2-dimensional boundary vanishes (pairwise cancellation of faces) for the 3-dimensional boundary of a 4-dimensional cube.

This construction guarantees conservation. Thus every 2-face of one 3-cube will give a contribution which is equal in magnitude and opposite in sign to that of the abutting 2-face of another one of the eight 3-cubes.

Elie Cartan seems to have been the first to apply $\partial\partial = 0$ to general relativity. He, Kibble, Sciama and Hehl have in this connection considered generalizations of Einstein's theory which call upon the geometrical notion of torsion as well as curvature. However, no one has yet proposed an experiment adequate in sensitivity to distinguish theories with torsion from Einstein's standard 1915 torsion-free theory.

Given curvature as our only building material and given one face of a 3-cube, we can calculate only the rotation associated with that face. That rotation is no good for constructing the energy momentum content of the 3-cube. (1) Added up for all six faces of the 3-cube it always gives a zero sum. (2) It is the wrong kind of geometric quantity. It has two indices rather than one. We turn to elementary statics for an idea (Figure 11).

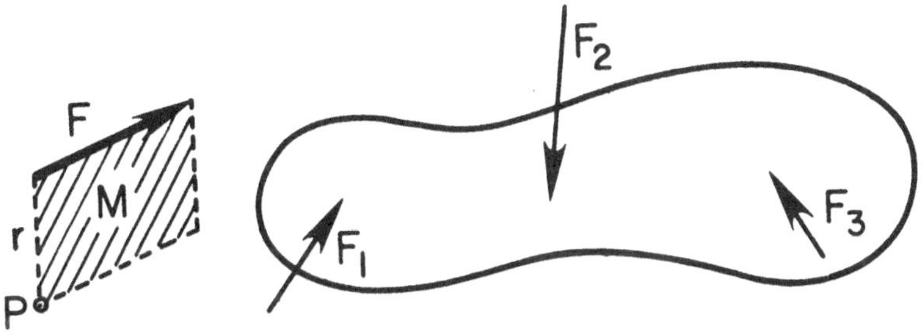

Figure 11. Forces and moments (prefiguring rotations and moments of rotation). Left, force F, and its moment, $M = r \times F$, with respect to the point P. Right, object in equilibrium under the action of F_i. The vector sum of the forces is zero, $F_i = 0$. Therefore, the sum of the moments of these forces about a given point, $M = \Sigma r_i \times F_i$, is independent of any alteration, a, in the location of that point: $M = \Sigma r_i \times F_i + 0 = \Sigma r_i \times F_i + a \times \Sigma F_i = \Sigma(r_i + a) \times F_i$.

For a body in static equilibrium the vector sum of the forces on it has to vanish. This granted, the sum of the moments of these forces is independent of the arbitrary point about which we choose to calculate this moment:

$$M = \Sigma(r_i + a) \times F_i = \Sigma r_i \times F_i. \tag{19}$$

The Concept of Moment of Rotation

Guided by the example of statics, we select some arbitrary point, P, inside or near the 3-cube we are considering. We take the distance, r_i, from that point to the center of the ith face of the 3-cube. We evaluate the "moment of rotation", $r_i \times (\text{rotation})_i$, associated with that face. We add this moment of rotation up for all six faces. We identify this with the 8π times the amount of energy momentum 4-vector in that 3-cube.

The sum of the rotations for the six faces is zero, as we have already seen. Therefore the sum of the moments of these rotations is independent of the arbitrary point about which we choose to calculate the moments. Moreover, this moment of rotation is a two-plus-one-equals-three-index quantity. It is "dual" (in the sense of 4-dimensional geometry) to a one-index quantity, a vector. The vector representation of moment of rotation we identify (up to the conventional factor 8π) with energy-momentum because it is automatically conserved. That is how we make our source, energy-momentum, the slave of our field, spacetime geometry:

$$\left\{\begin{array}{l}\text{dual of}\\ \text{moment of}\\ \text{rotation summed}\\ \text{over the 6 faces}\\ \text{of 3-cube}\end{array}\right\} = 8\pi\left\{\begin{array}{l}\text{amount of}\\ \text{momentum-energy}\\ \text{in that 3-cube}\end{array}\right\}. \qquad (20)$$

This is the true content and meaning of Einstein's field equation.

We can spell out this concept of moment of rotation more fully for any of the eight 3-cubes that bound our 4-cube. However, it is simplest to take as example a time-directed region, a 3-cube of dimension $\Delta x\ \Delta y, \Delta z$ centered at the point xyzt (all components of $d^3\Sigma_\beta$ zero except $d^3\Sigma_0$) and to inquire after only one component, the energy or time component, of what is contained in this region,

$$8\pi T_0{}^0 = \left\{\begin{array}{l}\text{dual of}\\ \text{moment of}\\ \text{rotation summed}\\ \text{over the 6 faces}\\ \text{of the 3-cube}\end{array}\right\}. \qquad (21)$$

To ask for the time component alone of the energy-momentum is to ask only for the xyz component of its dual, the moment of rotation itself. That means that we can disregard the time components of any of the six rotations. It is enough by way of illustration to look at the contribution of two opposing faces of the 3-cube. We take the arbitrary point P at the center of the 3-cube. We represent the moment of rotation as a parallelepiped (Figure 12).

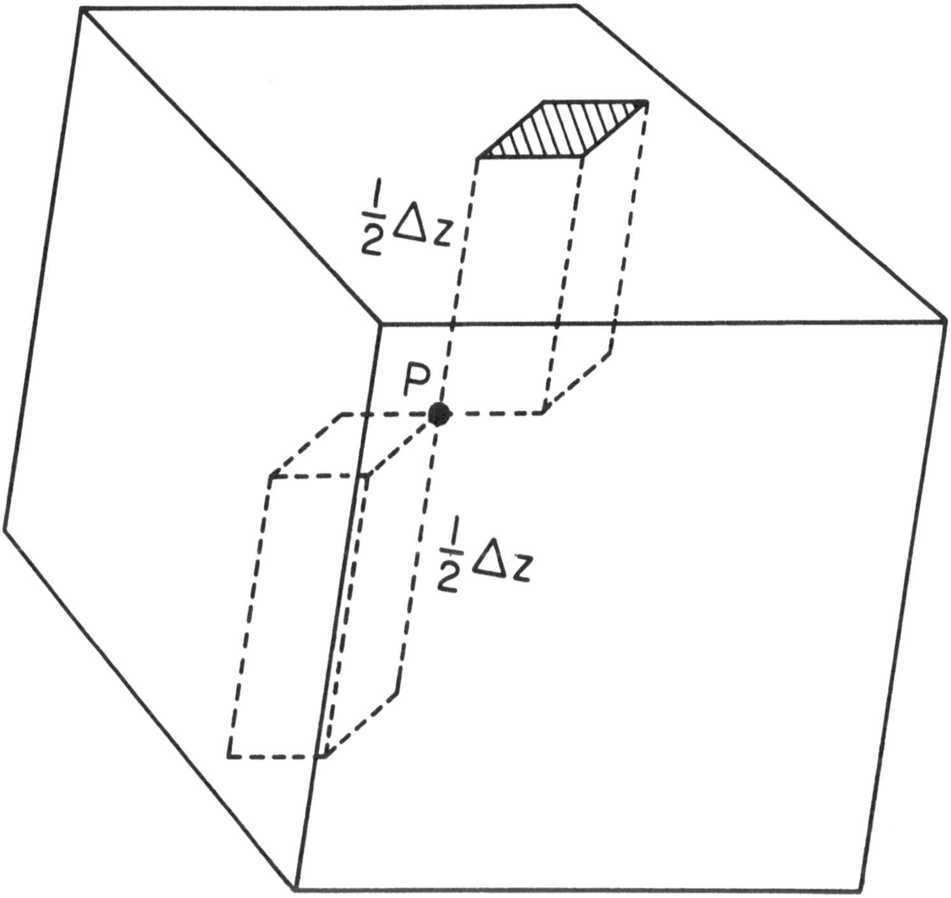

Figure 12. Moments of rotation associated with the top and bottom faces of the 3-cube $\Delta x\Delta y\Delta z$, represented as parallelepipeds. The combined 3-volume of these two parallelepipeds is independent of the location of the point P. The sum of the volumes of the parallelepipeds associated with all six faces tells how much energy there is in the 3-cube.

The moment of rotation associated with the top face of the 3-cube is:

$$\begin{pmatrix}\text{moment of}\\ \text{rotation}\end{pmatrix}_{\text{top face}} = -\underbrace{(\underbrace{R^{12}}_{\substack{\text{plane}\\ \text{of}\\ \text{rotation}}}{}_{12}\,.\underbrace{\Delta x\Delta y}_{\text{area}})}_{\text{rotation}}\ \underbrace{\frac{\Delta z}{2}}_{\substack{\text{lever}\\ \text{arm}}}\ . \tag{22}$$

When we add these moments of rotation for all six faces and divide by the volume of the 3-cube we get the moment of rotation per unit volume, and at last the physical interpretation of the Einstein tensor,

$$G_0{}^0 = \begin{Bmatrix} \text{moment of} \\ \text{rotation per} \\ \text{unit volume} \end{Bmatrix} = -(R^{12}{}_{12} + R^{23}{}_{23} + R^{31}{}_{31}). \qquad (23)$$

Duality

If only the principle $\partial\partial = 0$ came into play we might say that we had got everything out of nothing. However, we needed more, duality, which in essence means the idea of metric itself. Without that we could not have obtained the one-index vector of energy-momentum out of the three-index moment of rotation. That means that there are important elements in physics which still elude the principle of austerity.

To use the statement $\partial\partial = 0$ allows us also to deduce, even more simply, almost all of classical electrodynamics from almost nothing; and with a few more complications, almost all of chromodynamics from almost nothing. The almost trivial identity of algebraic geometry, that the boundary of a boundary is zero, pervades the breadth and depth of the field physics of our day. This circumstance gives us renewed occasion to take seriously the austerity hypothesis, that almost all physics is built on almost nothing.

The Continuing Search for the Origin of the Quantum Principle

Above all laws, all Lagrangians and all Hamiltonians stands the quantum, regulating principle of twentieth century physics. It tells us what are the allowable specifications of the state of the system. It reveals which measurements will have definite outcomes. It forecasts for other experimental set-ups the probability distribution of outcomes, the correlations between one quantity and another, and those products of uncertainties that define indeterminism.

Incontestable as quantum theory is, battle-tested as its predictions are, it still sometimes appears to us as something strange, unwelcome, and forced on us as it were from outside against our will, whereas if we really understood its deeper foundations, it would seem the most natural of all features of the world. We would be able to state the concept in a single clear sentence. We formulate special relativity in the words, "Physics looks the same in every free float frame of reference" and derive from that one idea a dozen sharp mathematical consequences. All the great formalism of quantum theory will surely someday unfold with equal ease out of some utterly elementary idea. Then we will all say to each other, "How stupid we were! Why didn't we see it before? The central idea is so simple, so natural, so beautiful that nobody could even imagine anything else. How could the world possibly be otherwise?"

We are far today from seeing the deeper lesson of the quantum or how it ties into the theme of "*austerity* ", of "almost everything from almost nothing". Nevertheless, we perhaps catch a glimmer of where to look from recent work of William Wootters on "statistical distinguishability" (Wootters, 1980, 1981). He goes bach to R.A. Fisher's analysis of the distinguishability of populations. In one population, for example, the proportion of blue eyes, grey eyes and brown eyes is p_1, p_2, p_3; and in a second population it is p_1', p_2', p_3'. How many individuals must one seize out of an invading force to know with some specified degree of assurance--say 10 to 1 odds--that the invaders belong to population I rather than II? What is the appropriate measure of statistical distinguishability in this sense of distinguishability, after one has divided out the obvious factor of the square root of the size of the captured sample? One might be tempted to represent two nearly alike populations as two nearby points in Euclidean (p_1, p_2, p_3) space and measure the distance between these two points in the plane triangle $p_1 + p_2 + p_3 = 1$. However, this measure of distinguishability "doesn't work". Populations I and II may be as distinguishable as populations III and IV ("same number of captives have to be taken to distinguish I from II as to distinguish III from IV") and yet the distance I-II, in one part of the plane, may be very different from the distance III-IV in another part of the plane.

Give up the p_1, p_2, p_3 diagram. Instead represent each population as a point in a $p_1^{1/2}$, $p_2^{1/2}$, $p_3^{1/2}$ diagram. Then each population lies on a curved surface, one-eighth of a sphere. In this diagram arc distance on the sphere between nearby points at last does measure statistical distinguishability.

The very task of distinguishing one population from another by statistical count thus drives one out of probabilities themselves and into probability amplitudes. Moreover, the considerations that apply to a population are equally relevant to a quantum state--for example, a quantum state described by three probability amplitudes ψ_1, ψ_2, ψ_3. Of course, there is an important distinction: the ψ_i are complex whereas the $p_i^{1/2}$ are real. Despite this distinction, Wootters is able to establish an important point: *angle in Hilbert space between two nearby quantum states is identical with the Wootters measure of the statistical distinguishability of those two states.*

We might not resonate so much to this result if we were not aware of the point emphasized with ever increasing force and clarity by Leibniz, Kant and Mach. All of what we call "reality", they remind us, goes back in the last analysis to the producing of order out of sense impressions: not first law and then sense impressions, but first sense impressions and out of them "law"; that is to say, regularities.

We find ourselves invited to ask at the end if the machinery of particles and fields, of spacetime and supersymmetry, is at bottom no machinery at all? Is this apparent machinery not built on something more subtle, more intangible than any machinery: on the statistics of something so humble as sense impressions? Is existence a gigantic information condensate?

Einstein asked if all the forces of nature could be unified in a single geometrical field. Heisenberg sought to build up all of particle physics out of a single non-linear ψ-field. In this meeting we have seen the marvellous progress being made in a still more sophisticated line of inquiry, generalizing older ideas of geometry into a new supersymmetry, and achieving many of the goals sought by both Einstein and Heisenberg.

Today we have less reason than ever before to let ourselves be driven away from the wonderful goal of *unity* that they held out before us.

Turning to the bust of Heisenberg

Heisenberg, we thank you for the insights you gave us
In the world of science.
We admire the spirit of hope
You brought to dispirited young people in dark days.
We think of you as an
Unforgettable representative
Of German learning and culture
In the great tradition.
Bravo!

Now let me thank the organizers of this meeting and the staff for the wonderful spirit that has accompanied us from beginning to end:

References

Bohr, N., and L. Rosenfeld, 1933, "Zur Frage der Messbarkeit der elektromagnetischen Feldgrößen," *Kgl. Danske Videnskab. Sels. Mat.-fys. Medd. 12*, no. 8. The English translation of this paper appears in *Selected Papers of Leon Rosenfeld*, R. Cohen and J. Stachel, eds., Reidel, Dordrecht, 1979.

Boltzmann, L., 1904, *Vorlesungen über die Prinzipien der Mechanik*, Vol. II, Wien. This paper and its intellectual antecedents are discussed in a recent paper by Dr. S. Wagner (1981) whom I thank for his communication on this point.

Christodoulou, D., 1970, "Reversible and irreversible transformations in black-hole physics," *Phys. Rev. Lett. 25*, 1596-1597.

Cordero, P., and C. Teitelboim, 1978, "Remarks on supersymmetric black holes," *Phys. Lett. 78B*, 80-83.

Cordero, P., post 1978, "On supersymmetric black holes," Proceedings of the First Chilean Symposium of Theoretical Physics, December 1978, special issue of *Contribuciones Cientificas y technologicas* (Universidad Tecnica del Estado, Avda. Ecuador 3469, Santiago, Chile), no date, pp.66-74.

Deser, S., and B. Zumino, 1976, "Consistent supergravity," *Phys. Lett. 62B*, 335-337.

Dirac, P.A.M., 1928a, "The quantum theory of the electron, I," *Proc. Roy. Soc. (London) A117*, 610-624.

Dirac, P.A.M., 1928b, "The quantum theory of the electron, II," *Proc. Roy. Soc. (London) A118*, 351-361.

Einstein, A., 1913, letter to Ernst Mach dated June 26, 1913, together with other correspondence between Einstein and Mach in F. Herneck, *Einstein und sein Weltbild*, Der Morgen, Berlin DDR, 1979.

Einstein, A., 1934, *Essays in Science*, Philosophical Library, New York. Translated from *Mein Weltbild*, Querido Verlag, Amsterdam, 1933.

Einstein, A., 1949, "Autobiographical Notes", in P.A. Schilpp, ed., *Albert Einstein : Philosopher-Scientist*, Library of Living Philosophers, Evanston, Ill., pp.65-67.

Feynamn, R.P., 1981, "Qualitative behavior of quantum chromodynamics," report at 3-5 April, 1981 Conference "The Way Ahead," University of Texas, Austin.

Freedman, D.L., P. van Nieuwenhuizen and S. Ferrara, 1976, "Progress toward a theory of supergravity," *Phys. Rev. D13*, 3214-3218.

Gibbons, G.W. and S.W. Hawking, 1977, "Action integrals and partition functions in quantum gravity," *Phys. Rev. D15*, 2752-2757.

Heisenberg, W., 1969, *Der Teil und das Ganze. Gespräche im Umkreis der Atomphysik*, (Munich), English translation, *Physics and Beyond. Memories of a Life in Science*, London, 1971.

Hojman, S., K. Kuchar and C. Teiteboim, 1973, "New approach to general relativity," *Nature Phys. Sci. 245*, 97-98.

Hojman, S., K. Kuchar and C. Teitelboim, 1976, "Geometrodynamics regained," *Ann. of Phys.* *76*, 88-135.

Isenberg, J., and J.A. Wheeler, 1979, "Inertia here is fixed by mass-energy there in every W model universe," in M. Pantaleo and F. deFinis, eds., *Relativity, Quanta and Cosmology in the Development of the Scientific Thought of Albert Einstein*, Johnson Reprint Corp., New York, Vol. I, pp. 267-293.

Jammer, M., 1974, *The Philosophy of Quantum Mechanics*, Wiley, New York.

Johnson, K., 1981, "The QCD vacuum," Report at the 16-21 July 1981 Heisenberg Symposium, Munich.

Kuchar, K., 1973, "Canonical quantization of gravity," pp. 238-288 in W. Israel, ed., *Relativity, Astrophysics and Cosmology*, Reidel Dordrecht, Holland.

Kuchar, K., 1974, "Geometrics regained: a Lagrangian approach," *J. Math. Phys.* *15*, 708-715.

Peres, A., 1962, "On tne Cauchy problem in general relativity, II," *Nuovo Cimento* *26*, , 53-62. The meaning of this equation is disucssed in Wheeler (1968).

Perry, M.J., S.W. Hawking and G.W. Gibbons, 1978, "Path integrals and indefiniteness of the gravitational action," *Nucl. Phys.* *B138*, 141-150.

Tabensky, R., and C. Teitelboim, 1977, "The square root of general relativity," *Phys. Lett.* *69B*, 453-456.

Teitelboim, C., 1973, "How commutators of constraints reflect space-time structure," *Ann. of Phys.* *79*, 542-557.

Teitelboim, C., 1973, "The Hamiltonian structure of spacetime", doctoral dissertation, unpublished, Princeton University; available from University Microfilms, Inc., Ann Arbor, Michigan 48106.

Teitelboim, C., 1976, "Surface deformations, spacetime structure and gauge invariance," in C. Aragone, ed., *Relativity, Fields, Strings, and Gravity: Proceedings of the Second Latin American Symposium on Relativity and Gravitarion SILARG II held in Caracas, December 1975* Universidad Simon Bolivar, Caracas.

Teitelboim, C., 1977, "Supergravity and square roots of constraints," *Phys. Rev. Lett.* *38*, 1106-1110.

Tomonaga, S., 1946, "On a relativistically invariant formulation of the quantum theory of wave fields," *Prog. Theor. Phys.* *1*, 27-42, reprinted in Schwinger, J., ed., *Selected Papers on Quantum Dynamics* Dover, New York, 1958.

Wagner, S., 1981, "Ludwig Boltzmann und Einstein's Relativitäts-theorien," contribution to 4 September 1981 Int. Conf. on Ludwig Boltzmann.

Weyl, H., 1949, *Philosophy of Mathematics and Natural Science* (original German in 1927; revised for English edition; translated by O. Helmer), Princeton University Press, Princeton, N.J., p. 91.

Wheeler, J.A.., 1957, "On the nature of quantum geometrodynamics", *Annals of Physics* , 604-614.

Wheeler, J.A., 1968, "Superspace and quantum geometrodynamics" in C.M. DeWitt and J.A. Wheeler, eds., *Battelle Rencontres*, Benjamin, New York, pp. 242-307.

Wheeler, J.A., 1978, "The 'past' and the 'delayed choice' double-slit experiment," in A.R. Marlow, ed., *Mathematical Foundations of Quantum Theory*, Academic Press, New York, pp. 9-48.

Wheeler, J.A., 1980, "Beyond the black hole," in H. Woolf, ed., *Some Strangeness in the Proportion: A Centennial Symposium to Celebrate the Achievements of Albert Einstein*, Addison-Wesley, Reading, Mass., pp. 341-375.

Wheeler, J.A., 1981, "Delayed-choice experiments and the Bohr-Einstein dialog," *The American Philosophical Society and the Royal Society: Papers Read at a Meeting June 5, 1980*, Philadelphia, pp. 9-40. The use of the double quasistellar object ("quasar" red shift z = 1.41) as setting for a delayed-choice experiment is described in figures 5 and 6.

Wootters, W.K., 1980, "The acquisition of information from quantum measurements," Ph.D. dissertation, University of Texas, Austin; available from University Microfilms, Inc., Ann Arbor, Mich. 48106.

Wootters, W.K., 1981, "Statistical distribution and Hilbert space," *Phys. Rev. D 23*, 357.

Lecture Notes in Physics

Vol. 136: The Role of Coherent Structures in Modelling Turbulence and Mixing. Proceedings 1980. Edited by J. Jimenez. XIII, 393 pages. 1981.

Vol. 137: From Collective States to Quarks in Nuclei. Edited by H. Arenhövel and A. M. Saruis. VII, 414 pages. 1981.

Vol. 138: The Many-Body Problem. Proceedings 1980. Edited by R. Guardiola and J. Ros. V, 374 pages. 1981.

Vol. 139: H. D. Doebner, Differential Geometric Methods in Mathematical Physics. Proceedings 1981. VII, 329 pages. 1981.

Vol. 140: P. Kramer, M. Saraceno, Geometry of the Time-Dependent Variational Principle in Quantum Mechanics. IV, 98 pages. 1981.

Vol. 141: Seventh International Conference on Numerical Methods in Fluid Dynamics. Proceedings. Edited by W. C. Reynolds and R. W. MacCormack. VIII, 485 pages. 1981.

Vol. 142: Recent Progress in Many-Body Theories. Proceedings. Edited by J. G. Zabolitzky, M. de Llano, M. Fortes and J. W. Clark. VIII, 479 pages. 1981.

Vol. 143: Present Status and Aims of Quantum Electrodynamics. Proceedings, 1980. Edited by G. Gräff, E. Klempt and G. Werth. VI, 302 pages. 1981.

Vol. 144: Topics in Nuclear Physics I. A Comprehensive Review of Recent Developments. Edited by T.T.S. Kuo and S.S.M. Wong. XX, 567 pages. 1981.

Vol. 145: Topics in Nuclear Physics II. A Comprehensive Review of Recent Developments. Proceedings 1980/81. Edited by T. T. S. Kuo and S. S. M. Wong. VIII, 571-1.082 pages. 1981.

Vol. 146: B. J. West, On the Simpler Aspects of Nonlinear Fluctuating. Deep Gravity Waves. VI, 341 pages. 1981.

Vol. 147: J. Messer, Temperature Dependent Thomas-Fermi Theory. IX, 131 pages. 1981.

Vol. 148: Advances in Fluid Mechanics. Proceedings, 1980. Edited by E. Krause. VII, 361 pages. 1981.

Vol. 149: Disordered Systems and Localization. Proceedings, 1981. Edited by C. Castellani, C. Castro, and L. Peliti. XII, 308 pages. 1981.

Vol. 150: N. Straumann, Allgemeine Relativitätstheorie und relativistische Astrophysik. VII, 418 Seiten. 1981.

Vol. 151: Integrable Quantum Field Theory. Proceedings, 1981. Edited by J. Hietarinta and C. Montonen. V, 251 pages. 1982.

Vol. 152: Physics of Narrow Gap Semiconductors. Proceedings, 1981. Edited by E. Gornik, H. Heinrich and L. Palmetshofer. XIII, 485 pages. 1982.

Vol. 153: Mathematical Problems in Theoretical Physics. Proceedings, 1981. Edited by R. Schrader, R. Seiler, and D.A. Uhlenbrock. XII, 429 pages. 1982.

Vol. 154: Macroscopic Properties of Disordered Media. Proceedings, 1981. Edited by R. Burridge, S. Childress, and G. Papanicolaou. VII, 307 pages. 1982.

Vol. 155: Quantum Optics. Proceedings, 1981. Edited by C.A. Engelbrecht. VIII, 329 pages. 1982.

Vol. 156: Resonances in Heavy Ion Reactions. Proceedings, 1981. Edited by K.A. Eberhard. XII, 448 pages. 1982.

Vol. 157: P. Niyogi, Integral Equation Method in Transonic Flow. XI, 189 pages. 1982.

Vol. 158: Dynamics of Nuclear Fission and Related Collective Phenomena. Proceedings, 1981. Edited by P. David, T. Mayer-Kuckuk, and A. van der Woude. X, 462 pages. 1982.

Vol. 159: E. Seiler, Gauge Theories as a Problem of Constructive Quantum Field Theory and Statistical Mechanics. V, 192 pages. 1982.

Vol. 160: Unified Theories of Elementary Particles. Critical Assessment and Prospects. Proceedings, 1981. Edited by P. Breitenlohner and H.P. Dürr. VI, 217 pages. 1982.

Selected Issues from

Lecture Notes in Mathematics

Vol. 728: Non-Commutative Harmonic Analysis. Proceedings, 1978. Edited by J. Carmona and M. Vergne. V, 244 pages. 1979.

Vol. 729: Ergodic Theory. Proceedings 1978. Edited by M. Denker and K. Jacobs. XII, 209 pages. 1979.

Vol. 730: Functional Differential Equations and Approximation of Fixed Points. Proceedings, 1978. Edited by H.-O. Peitgen and H.-O. Walther. XV, 503 pages. 1979.

Vol. 731: Y. Nakagami and M. Takesaki, Duality for Crossed Products of von Neumann Algebras. IX, 139 pages. 1979.

Vol. 733: F. Bloom, Modern Differential Geometric Techniques in the Theory of Continuous Distributions of Dislocations. XII, 206 pages. 1979.

Vol. 735: B. Aupetit, Propriétés Spectrales des Algèbres de Banach. XII, 192 pages. 1979.

Vol. 738: P. E. Conner, Differentiable Periodic Maps. 2nd edition, IV, 181 pages. 1979.

Vol. 742: K. Clancey, Seminormal Operators. VII, 125 pages. 1979.

Vol. 755: Global Analysis. Proceedings, 1978. Edited by M. Grmela and J. E. Marsden. VII, 377 pages. 1979.

Vol. 756: H. O. Cordes, Elliptic Pseudo-Differential Operators – An Abstract Theory. IX, 331 pages. 1979.

Vol. 760: H.-O. Georgii, Canonical Gibbs Measures. VIII, 190 pages. 1979.

Vol. 762: D. H. Sattinger, Group Theoretic Methods in Bifurcation Theory. V, 241 pages. 1979.

Vol. 765: Padé Approximation and its Applications. Proceedings, 1979. Edited by L. Wuytack. VI, 392 pages. 1979.

Vol. 766: T. tom Dieck, Transformation Groups and Representation Theory. VIII, 309 pages. 1979.

Vol. 771: Approximation Methods for Navier-Stokes Problems. Proceedings, 1979. Edited by R. Rautmann. XVI, 581 pages. 1980.

Vol. 773: Numerical Analysis. Proceedings, 1979. Edited by G. A. Watson. X, 184 pages. 1980.

Vol. 775: Geometric Methods in Mathematical Physics. Proceedings, 1979. Edited by G. Kaiser and J. E. Marsden. VII, 257 pages. 1980.

Vol. 779: Euclidean Harmonic Analysis. Proceedings, 1979. Edited by J. J. Benedetto. III, 177 pages. 1980.

Vol. 780: L. Schwartz, Semi-Martingales sur des Variétés, et Martingales Conformes sur des Variétés Analytiques Complexes. XV, 132 pages. 1980.

Vol. 782: Bifurcation and Nonlinear Eigenvalue Problems. Proceedings, 1978. Edited by C. Bardos, J. M. Lasry and M. Schatzman. VIII, 296 pages. 1980.

Vol. 783: A. Dinghas, Wertverteilung meromorpher Funktionen in ein- und mehrfach zusammenhängenden Gebieten. Edited by R. Nevanlinna and C. Andreian Cazacu. XIII, 145 pages. 1980.

Vol. 786: I. J. Maddox, Infinite Matrices of Operators. V, 122 pages. 1980.

Vol. 787: Potential Theory, Copenhagen 1979. Proceedings, 1979. Edited by C. Berg, G. Forst and B. Fuglede. VIII, 319 pages. 1980.

Vol. 791: K. W. Bauer and S. Ruscheweyh, Differential Operators for Partial Differential Equations and Function Theoretic Applications. V, 258 pages. 1980.

Vol. 792: Geometry and Differential Geometry. Proceedings, 1979. Edited by R. Artzy and I. Vaisman. VI, 443 pages. 1980.

Vol. 793: J. Renault, A Groupoid Approach to C*-Algebras. III, 160 pages. 1980.

Vol. 798: Analytic Functions, Kozubnik 1979. Proceedings. Edited by J. Ławrynowicz. X, 476 pages. 1980.

Vol. 799: Functional Differential Equations and Bifurcation. Proceedings 1979. Edited by A. F. Izé. XXII, 409 pages. 1980.

Vol. 801: K. Floret, Weakly Compact Sets. VII, 123 pages. 1980.

Vol. 802: J. Bair, R. Fourneau, Etude Géometrique des Espaces Vectoriels II. VII, 283 pages. 1980.

Vol. 804: M. Matsuda, First Order Algebraic Differential Equations. VII, 111 pages. 1980.

Vol. 805: O. Kowalski, Generalized Symmetric Spaces. XII, 187 pages. 1980.

Vol. 807: Fonctions de Plusieurs Variables Complexes IV. Proceedings, 1979. Edited by F. Norguet. IX, 198 pages. 1980.

Vol. 809: I. Gumowski and Ch. Mira, Recurrences and Discrete Dynamic Systems. VI, 272 pages. 1980.

Vol. 810: Geometrical Approaches to Differential Equations. Proceedings 1979. Edited by R. Martini. VII, 339 pages. 1980.

Vol. 816: L. Stoica, Local Operators and Markov Processes. VIII, 104 pages. 1980.

Vol. 819: Global Theory of Dynamical Systems. Proceedings, 1979. Edited by Z. Nitecki and C. Robinson. IX, 499 pages. 1980.

Vol. 821: Statistique non Paramétrique Asymptotique. Proceedings, 1979. Edited by J.-P. Raoult. VII, 175 pages. 1980.

Vol. 826: Ph. G. Ciarlet et P. Rabier, Les Equations de von Kármán. VI, 181 pages. 1980.

Vol. 827: Ordinary and Partial Differential Equations. Proceedings, 1978. Edited by W. N. Everitt. XVI, 271 pages. 1980.

Vol. 831: Representation Theory I. Proceedings, 1979. Edited by V. Dlab and P. Gabriel. XIV, 373 pages. 1980.

Vol. 832: Representation Theory II. Proceedings, 1979. Edited by V. Dlab and P. Gabriel. XIV, 673 pages. 1980.

Vol. 836: Differential Geometrical Methods in Mathematical Physics. Proceedings, 1979. Edited by P. L. García, A. Pérez-Rendón, and J. M. Souriau. XII, 538 pages. 1980.

Vol. 837: J. Meixner, F. W. Schäfke and G. Wolf, Mathieu Functions and Spheroidal Functions and their Mathematical Foundations Further Studies. VII, 126 pages. 1980.

Vol. 838: Global Differential Geometry and Global Analysis. Proceedings 1979. Edited by D. Ferus et al. XI, 299 pages. 1981.

Vol. 840: D. Henry, Geometric Theory of Semilinear Parabolic Equations. IV, 348 pages. 1981.

Vol. 841: A. Haraux, Nonlinear Evolution Equations- Global Behaviour of Solutions. XII, 313 pages. 1981.

Vol. 842: Séminaire Bourbaki vol. 1979/80. Exposés 543–560. IV, 317 pages. 1981.

Vol. 843: Functional Analysis, Holomorphy, and Approximation Theory. Proceedings. Edited by S. Machado. VI, 636 pages. 1981.

Vol. 845: A. Tannenbaum, Invariance and System Theory: Algebraic and Geometric Aspects. X, 161 pages. 1981.

Vol. 849: P. Major, Multiple Wiener-Itô Integrals. VII, 127 pages. 1981.